The End
of the Long Summer

The End
of the Long Summer

WHY WE MUST REMAKE OUR CIVILIZATION
TO SURVIVE ON A VOLATILE EARTH

Dianne Dumanoski

THREE RIVERS PRESS
NEW YORK

Published in the United States by Three Rivers Press, an imprint of the
Crown Publishing Group, a division of Random House, Inc., New York.

www.crownpublishing.com

THREE RIVERS PRESS and the Tugboat design are registered trademarks of
Random House, Inc.

Originally published in hardcover in the United States by Crown Publishers,
an imprint of the Crown Publishing Group,
a division of Random House, Inc., New York, in 2009.

Library of Congress Cataloging-in-Publication Data is available upon request

ISBN 978-0-307-39609-9

Printed in the United States of America

Design by Leonard W. Henderson

10 9 8 7 6 5 4 3 2 1

First Paperback Edition

Contents

The End
of the Long Summer

1

The Future Head-On

The future in the modern imagination has always stretched out ahead like a broad highway drawing us onward with the promise of tomorrow. Now rather suddenly, as it becomes impossible to ignore dramatic physical changes taking place across the Earth, the future looms like an urgent question. Whatever the coming century brings, it will not unfold smoothly as some improved but largely familiar version of life as we know it. This is the only thing that seems certain.

Now that many are beginning to awake to the danger, perhaps we will finally come to understand the stakes in the planetary emergency unfolding all around us—an emergency that involves far more than the pressing problem of climate change. I've long had a bumper sticker pinned to my office door that captures the truth not yet fully grasped. It bears a blue cartoon-like outline of a sperm whale along with the admonition: SAVE THE HUMANS. Whatever else is in jeopardy, this is first and foremost a crisis for humans and our current civilization. Moreover, what confronts us is not a vague prospective danger to an abstract posterity in some future time. The threat is to a child born today.

In recent years, scientists have been drilling into the ice sheets on Greenland and Antarctica and retrieving cores of ice, some almost two miles long, which contain trapped bubbles of ancient air—bubbles that record the long history of past conditions on Earth over some 800,000 years and also hold clues to the future. The deep ice tells us many things, but two facts about our situation stand out. First, human activity is pushing the balance of gases in the Earth's atmosphere way beyond the normal range at an alarming speed. Second, the world as we know it with agriculture, civilization, and dense human numbers emerged during a rare interlude of climatic grace—a "long summer" of unusual stability over the past 11,700 years. Because of humanity's planetary impact, this benign period is now ending.

Today, the hyperactive growth of industrial capitalism and the burden of increasing human numbers constitute a *planetary* force comparable in disruptive power to the ice ages and asteroid collisions that have previously redirected Earth's history. Since the first Earth Day in 1970 and the 1972 global conference on the environment in Stockholm, there has been growing alarm about the ways human activity everywhere around the globe undermines the renewal powers that sustain local and regional ecosystems. More recently, however, scientists began reporting ominous developments heralding dangers of a whole new order. In the second half of the twentieth century, the relentless expansion of modern industrial civilization started to impinge on the invisible, global-scale cycles that make up the Earth's essential life support. The human enterprise had become an agent of risky *global* change that now threatened to undermine fundamental parts of the Earth's metabolism.

In the ultimate irony, however, human domination of the Earth has not brought with it the control of nature promised by the modern era's guiding myth of progress. Nor has it brought "the end of nature" as the author of an early book on global warming lamented. Rather we are already witnessing Nature's return to center stage as a critical player in human history. This development, more than any other, will shape the human future.

The urgent question is, simply stated, whether in the face of these changes, the Earth will remain a place that can support complex, interconnected global civilization or, in the extreme, sustain human life.

Over the past two decades, attention has focused on specific symptoms of humanity's global-scale disruption of the environment: the destruction of the ozone layer, climate change, the worldwide loss of species, the growing threats to oceans, pervasive chemical contamination of food webs everywhere on Earth. But these are only signs of broader planetary distress, aspects of a larger story. We need to press beyond the symptoms to discover what fundamentally ails us and to gain a clear-eyed view of where we now find ourselves.

The End of the Long Summer looks anew at the human story and sets forth an account radically different from the onward and upward progress narrative of the modern era. The source of its hope lies not in the belief that humans are destined to achieve dominion but rather in the evidence that we are a stormworthy lineage that has managed to flourish on an increasingly volatile Earth. We come from a long line of survivors who were tempered in the crucible of climatic reversals and catastrophic change. This book also explores the challenge of living in a time of great uncertainty—a challenge our forebears faced repeatedly in their evolutionary passage—and what this moment requires of us. Above all else, it concerns "the obligation to endure." Biologist Jean Rostand's resonant phrase, famously quoted by Rachel Carson in *Silent Spring*, asks us to see ourselves in a longer perspective than our own lives; it reminds us of our responsibility in life's compact across generations to those who preceded us and to those who will follow. It is time to confront our dilemma squarely and learn how and why we have arrived at this perilous juncture. We need to understand what we are up against if we are to make wise decisions about how to proceed in a time of growing danger.

In my work as a journalist, I spent more than a quarter century on the front lines of the crisis in nature, witnessing firsthand the destructiveness of the modern human enterprise. I've served as a chronicler of loss and a messenger of warnings over the time that this crisis has escalated from the early concerns about dirty air and dirty water of that first Earth Day to the current global jeopardy of deteriorating and unstable planetary systems. As the situation worsened, those trying to remedy the multiplying symptoms of stress have hurtled from one emergency to another, running faster and faster on an accelerating treadmill of crisis. Treating symptoms has been akin to chasing brush fires. Finally there is no choice but to take on the pyromaniac.

Our dilemma is more than an "environmental crisis." The modern era has been a radical cultural experiment. Without question, it has been spectacularly successful in the short run at producing wealth and comfort for more people than ever before. But this success entails a dangerous gamble. The global civilization that now dominates the world has departed fundamentally from practices that have helped ensure human survival in the longer run.

If we are to come to grips with this planetary emergency, we need to understand the process that drives it. Most of all, we need to see how our current modes of thinking fuel this emergency and at the same time increase our vulnerability to the consequences. The inherited assumptions we bring to the situation impair our assessment of the dangers. In this new historical landscape, we not only continue to intensify the physical crisis through exponential growth. Perhaps more important, we also struggle to understand and resolve our dilemma using ideas about the world that are now obsolete and dangerous.

This failure of fundamental ideas lies at the heart of the broader human crisis. Modern industrial civilization's recent capacity to disrupt essential planetary systems has precipitated stunning events, such as the appearance of the Antarctic ozone hole, that challenge our culture's view of the world. The past two decades have raised fundamental questions about human power, the nature of the world we inhabit and act upon, and humanity's place in it. This broader crisis cannot be remedied through increased scientific research or short-term technological fixes. In this new era, our inherited cultural map lies in tatters and needs serious revision to serve as a reliable guide. The question of whether we correctly understand both our own presence on Earth and the nature we disrupt is not merely a philosophical concern. It bears directly on practical decisions that lie ahead and on the deep, longer-term changes needed to redirect the human enterprise onto a safer course. For four centuries, it did not seem to matter that the vast construct of modern civilization rested on an inaccurate view of nature or that our cultural map was missing vital information.

Now it matters most of all.

The problem of climate change illuminates the radical uncertainty of our situation and the uncomfortable fact that the future is not in our hands alone. Leading voices, including former UN chief weapons inspector Hans Blix, have warned that climate change will be a far greater hazard in the twenty-first century than terrorism. In truth, there is no knowing exactly how events will unfold. Scientists working with computer models of the Earth's climate system can warn about where we *may* be headed if levels of carbon dioxide in the atmosphere reach levels twice what they were at the start of the Industrial Revolution, as is likely by midcentury on our current business-as-usual trajectory. But the outcome will depend not only on *what we do* but also on *how Earth responds*. In the past few years, new scientific studies have indicated that the Earth's response is proving "faster and nastier," as one British policy specialist put it, than current climate models forecast. The *possible* future, outlined in the consensus scientific assessment compiled under the auspices of the United Nations, ranges from seriously disruptive to dangerous and even catastrophic.

The hard truth is that there are no "solutions" that can simply halt this planetary emergency, stop the dramatic changes that are already under way, and save the world we have known. In squandering the chance to avoid significant climate change, we've already crossed one fateful threshold. But other, much more dangerous thresholds lie ahead. With resolve and foresight, we can foster the resilience that is essential in a time of growing instability; with rapid, determined action as well as luck, we may perhaps avoid outright catastrophes of rapid sea-level rise, abrupt climatic jolts that would shake our civilization off its foundations, and irreversible changes that would make most of the planet inhospitable for human life. In the past decade, however, the trends in our global economy have been sweeping us like a riptide out into ever more dangerous waters.

The task then is to do our utmost to avoid the worst and, at the same time, figure out how to weather the change that is now unavoidable. *How* climate changes is going to be as important in coming decades as how much temperatures rise. Contrary to the common and persistent notion that global warming is going to proceed like a smooth escalator carrying northern climes into an era of balmy winters, scientists studying the Earth's early climate history have found evidence of swift, intense change in less than a century or even within a single decade. The faster the warming and the higher temperatures climb, the greater the danger that change will arrive in abrupt shifts and surprises—shocks that could lead to the collapse of social and economic systems. The dramatic loss of ozone was such a surprise.

Recent media coverage of global warming and Al Gore's film *An Inconvenient Truth* have highlighted the threat of melting ice sheets, rising seas, and the loss of low-lying areas and coastal cities as a worst-case scenario. Although catastrophic, these are relatively gradual changes to which we might, however chaotically and painfully, adapt. The real danger of abrupt shifts in planetary systems has figured little in the public discussion of global changes. Yet within the past 120,000 years, the climate in the North Atlantic region that embraces critical world financial centers in Europe and the United States has experienced repeated, extremely abrupt warming events as well as equally dramatic episodes of cooling. In some of these reversals, the climate has passed from one state to another as if by the flip of a switch in a decade or less, and *average* temperatures have shifted by as much as 18 degrees F.

The other pressing concern is whether the changes under way will increase climate fluctuations, bringing wild swings in temperature and rainfall. The climate during the long summer has been distinctly at odds with the chaotic climate humans have faced through most of our history, which often varied more *from decade to decade* than it has during the past twelve thousand years. A return to past patterns of extreme variability would be devastating to agriculture, which has provided the foundation for complex civilization over the past six millennia.

If, indeed, surprising shifts, abrupt change, and increasing climatic variability are among the possible challenges in our future, we must turn our attention to a new aim, as yet largely unconsidered—the task of shockproofing our human systems. The modern way of organizing life leaves us badly prepared for the disruption and instability it has engendered. The current trend toward interdependence and globalization is only increasing our vulnerability. From an evolutionary perspective, the process of globalization is a risky strategy indeed. Such tight integration may be an acceptable gamble if one lives in a relatively stable environment and can be reasonably confident that tomorrow will be pretty much like today. But it is not a wise way to meet a changing world. The pursuit of global integration at this time poses a real danger that critical institutions could unravel or suffer outright collapse from relatively minor climate discontinuities. We could, in short, find ourselves in the midst of social and economic chaos long before the advent of any climate catastrophe. This growing vulnerability of the global economic system has been largely absent from the discussion of climate change and the human future.

In the face of possibly severe challenges, we should move to restructure human society in ways that will make it less vulnerable to collapse in an unstable world, adopting into our human systems those features that have allowed Earth's life to endure in the face of catastrophe. Structural principles that have made the Earth system re-markably resilient include functional redundancy, in which a variety of different species do the same job; a modular structure in which the whole is comprised of smaller, relatively self-sufficient units; and com-partmentalization, which limits the connections of parts of the system to the whole. I am not suggesting a retreat from globalization alto-gether. Because of our planetary impact, all humans now share a com-mon destiny, which must be considered at the global scale and perhaps eventually managed through an effective global body. At the same time, however, the growing instability and uncertainty of our situation requires a fundamental shift in priorities. As we make choices about the future, we need to give serious thought to a new security founded on increased local and regional self-reliance rather than thinking sim-ply about efficiency and cost. The more we deglobalize and decentral-ize food and energy systems, the more likely we'll be able to meet our basic needs in an emergency when transportation becomes impossible. Given the foreseeable end of cheap oil in the coming decades and the political volatility of the Middle East, this makes good sense even if we weren't facing the challenges of global change.

There are other things to consider, as well. We must also find ways to enhance what some have called social capital—the capacity for trust and cooperation that helped our ancestors survive past calamities. It would also be prudent to develop simple and elegant es-sential technologies—for example, to purify water, generate electric-ity, and cook food—that can still operate and meet basic needs if the current global infrastructure breaks down, whether temporarily or for longer periods. And in the extreme, if it becomes clearer that our complex global civilization may not survive the chaos of the coming century intact, we must decide how to convey an essential cultural legacy to those who rebuild.

The aim in all this is *survivability*, a challenge that goes beyond "adapting" to a drier climate or flood dangers or making current practices "sustainable." First and foremost, we need to insulate and re-design our social and economic systems so they can better withstand disruption and shock and can change in the face of altered circumstance. Survivability should not be mistaken for survivalism—an impulse focused on retreat from society and individual survival. The aim rather is to safeguard the human knowledge and institutions that give us the *capacity* to respond with imagination and flexibility to a changing world.

The most formidable obstacle ahead may be an imaginative one. The first step is to recognize that we have entered a period of deep change. Of course, simply suggesting that our civilization may be hitting a dead end is considered a message of "gloom and doom." But this judgment is a matter of perspective. Acknowledging that we're at the end of something means we're at the start of something else. We need to imagine futures that don't much resemble the present—all kinds of futures, creative alternatives as well as frightening scenarios. The question is *not* how to preserve the status quo, but rather how to make our way in a new historical landscape. Today's children will likely confront challenges we can hardly begin to imagine in a radically altered, unrecognizable world. Can we responsibly continue preparing them for business as usual? And if not, what can we do to make them ready for a survival game in which wild cards rule?

This transition could span several volatile centuries, fraught with uncertainty, but humans have weathered radical changes before. Our lineage evolved on an erratic and inconstant Earth and not, as myths have imagined, in the tranquil security of some prehistoric Eden. Our evolutionary legacy has endowed us with the flexibility vicissitude requires and the creativity to survive in an uncertain world. We owe our very existence to ancestors who survived through times of cataclysmic change.

Now, in the years ahead, our children and grandchildren and their grandchildren must make a dangerous passage through a storm of our own making. The door to the comfortable and familiar world we depend on has already slammed shut behind us. It is already too late to "prevent" global warming or to "solve" the climate crisis, too late to prevent powerful forces from altering the trajectory of human history. That we have already crossed some ominous thresholds, however, does *not* mean that it is too late to do anything at all. We humans are at a critical juncture—an historical moment that requires courage and sober realism. We cannot bank on the end of the world or deliverance from the trials of existence, whether through biblical apocalypse or our own extinction. Nor can we proceed on blind faith in technological salvation. Fear, despair, and denial are indulgences we cannot afford. It is time to turn and face the future head-on.

The Planetary Era

We do not yet understand our own time.

As the twentieth century was drawing to a close, every news organization, it seemed, prepared some sort of retrospective highlighting the significant events and milestones of the century or of the past one thousand years. It was one of those rare moments when the media step back from the day-to-day hustle for a longer view. During the countdown to the new millennium, I found myself reading and watching these reviews with curiosity to see what was included and what left out. Then I began scouring them, searching for some sign that the leaders and communicators in our culture have even an inkling about where the twentieth century had taken us.

Two themes dominated these surveys of the departing century: the horrors of modern war and the triumphs of technology. In the replay of significant moments, Allied soldiers in World War II stormed onto the beach at Normandy, the fascinating and terrifying mushroom cloud surged skyward over Hiroshima, and soldiers liberating the Nazi extermination camps wandered amid surreal piles of pale, naked, bony bodies. Cold War highlights featured President John F. Kennedy's address to the nation during the Cuban missile crisis and shots of jubilant Germans astride the Berlin Wall, pulling it to pieces as the Soviet empire dissolved into history. Fragile single-engine airplanes lifted into tentative flight, and some of the first horseless carriages chugged across the TV screen, raising a rooster tail of dust.

A doctor in a white coat administered polio shots to children waiting in a long line. The linked ribbons of a double helix rotated on the screen, fraught with mystery and promise.

In nods to transforming social movements of the century, suffragettes marched for the vote beneath fluttering banners and magnificent broad hats, and Martin Luther King intoned "I have a dream." There were pictures of twitching, dying, DDT-poisoned birds, as viewers heard about writer Rachel Carson's enduring book *Silent Spring* and how this passionate indictment of the chemical war against nature had sparked the modern environmental movement.

But the image that transcended all others was that grainy shot of astronaut Neil Armstrong taking the first step onto the desolate landscape of the moon with the self-conscious pronouncement "That's one small step for a man, one giant leap for mankind." In the popular mind, this was the emblematic moment of the twentieth century.

In its final days, I watched Neil Armstrong step onto the moon again and again while I waited in vain to read or hear even a passing mention of the Antarctic ozone hole or recognition of the profound watershed in the human journey it symbolized—the arrival of a new and ominous epoch when human activity began to disrupt the essential but invisible planetary systems that sustain a dynamic, living Earth. Humanity had, indeed, taken a giant leap, but these retrospectives were seriously mistaken about the geography of the future. In the second half of the twentieth century, modern civilization emerged as a global-scale force capable of redirecting Earth's history. This fateful step marks a fundamental turning point in the relationship between humans and the Earth, arguably the biggest step since human mastery of fire, which helped launch the human career of dominion. The consequences are not limited to global warming, nor are weather extremes the first evidence of our new status. Accelerating climate change signals a far deeper problem—the growing human burden on *all* of the fundamental planetary processes that together make up a single, self-regulating Earth. When future historians look back on the twentieth century, this quick visit to the moon will surely seem like a minor event compared to the giant leap humanity had taken here on Earth.

The Earth is one in ways we are only beginning to understand.

Much more than merely an assemblage of ecosystems or a catalog of species, the Earth, not unlike the human body, is a dynamic whole that emerges from the interaction of all of life, the oceans, the air, the soil, and the rocks. It functions as a unified system with a global metabolism that depends on the living organisms that inhabit it—microbes, plants, and animals—as well as on chemical and geological processes, including the weathering of rocks, the eruption of volcanoes, and the downward plunge—subduction—of tectonic plates that make up the Earth's crust. This great global metabolism is what keeps the Earth a suitable place for life. Without this nonstop planetary maintenance, the Earth would not be the shimmering, inviting, cloud-draped blue and green orb we have only recently come to see as a whole in photographs astronauts have taken from space. Various parts of this system help maintain the balance of gases in the atmosphere, modulate the Earth's temperature, shield our planet from the sun's dangerous radiation, and recycle water and elements vital to life: carbon, nitrogen, sulfur, and phosphorous. In a great flux that developed over billions of years, these four nutrients, which are essential to life on the planet, constantly circulate in endless interconnected cycles far and wide across the face of the Earth as they move through plants and animals to soils, oceans, and atmosphere and back again to plants and animals.

The sulfur on land, for example, washes into rivers and is carried to the ocean, where it is taken up by one-celled marine plants, including some armored with tiny buttonlike shells called coccolithophores. One species of this group, *Emiliania huxleyi*—individuals measure one hundredth of a millimeter in diameter and look like a Hollywood designer's vision of a spherical space station that one might see in *Star Wars*—is abundant all over the world and can bloom with such exuberance that its presence is visible from space as milky swirls curling across great expanses of blue ocean. These tiny plants emit a sulfur-bearing gas, dimethyl sulfide, back into the atmosphere. There the sulfur particles may help regulate climate by reflecting incoming solar energy back to space and by attracting water droplets and stimulating the formation of clouds that help cool the earth. As winds blow these clouds from the ocean to land, the sulfur returns to earth with the rain and becomes available again to plants and other living things.

The well-being or ill health of any particular place on Earth depends on distant connections. Dust transported from the Sahel region of Africa provides essential elements for the great Amazon rainforest, and the native forests in Hawaii have endured on old, highly weathered volcanic rocks because the phosphorous needed to maintain them blows in with dust from the Gobi Desert more than 3,700 miles away.

This radical new view of life on Earth may prove to be the U.S. space program's greatest legacy. Perhaps ironically, the initiative that opened the door to this new perspective was the National Aeronautics and Space Administration's project to search for life on Mars. As this effort got under way in the early 1960s, a team at NASA's Jet Propulsion Laboratories in Pasadena, California, took up the question of how exactly one might discover whether Mars and other planets harbored life. One of the outside consultants advising on the design of instruments for this quest was a British scientist and inventor, James Lovelock, who had already gained some prominence in scientific circles for his invention of an exquisitely sensitive device—the electron capture detector—that could measure tiny amounts of chemical compounds in the atmosphere and elsewhere. In periodic visits to JPL, Lovelock quickly became engaged in the larger questions involved in this investigation. The initial thought was to design equipment that would land on

Mars to take samples and do tests in much the way one might investigate life on Earth. In time, however, Lovelock came to question this assumption. Perhaps if life on Mars existed, it would be life of a different style and not revealed by tests designed to study earthly life. And that led to even more basic and problematic questions, such as What is life? and How would we recognize it if we encountered life unlike ours?

Lovelock eventually hit on the idea that one could better recognize life elsewhere by looking for the signature of its *process* rather than for particular organisms. "Life," as one eminent biologist put it, "is a verb," and living things, in maintaining themselves through the incessant chemical activity and energy flow of metabolism, leave telltale signs of their presence in oceans and atmosphere. Since Mars has no ocean, the place to focus a life-detection experiment there would be its atmosphere.

Based on this theory, one would expect a planet without life to have a chemically static atmosphere, since it would have exhausted all the possible chemical reactions and arrived at equilibrium. The atmosphere of a life-bearing planet would be noticeably different. If there were life on Mars, living things would alter the atmosphere as they extracted nutrients for sustaining themselves and later released waste; the ongoing process of staying alive would move the atmosphere toward a state of disequilibrium. To the chagrin of some of his NASA colleagues bent on space exploration, such as astronomer Carl Sagan, Lovelock concluded that one did not need to send a spaceship to Mars to determine whether it had life. A telescope could detect this disequilibrium, the chemical fingerprint of life, even at a great distance. As early as 1965, in fact, readings from an infrared telescope at an observatory in France would provide a detailed analysis of the atmosphere on two relatively nearby neighbors in our solar system— Mars and Venus—readings that showed that both had atmospheres dominated by carbon dioxide and chemically close to equilibrium. Dead planets.

The atmosphere of Earth, by contrast, is an extraordinary mix of unstable gases that persists in a state of deep disequilibrium. Oxygen and methane do not by their nature remain long in the presence of each other because they react to form other, more stable compounds— carbon dioxide and water—yet they nevertheless coexist in predictable proportions. Similarly, the dominant gas in the Earth's atmosphere, nitrogen, was also a puzzle to Lovelock, for gaseous nitrogen is disposed to react with oxygen to form nitric acid, which eventually ends up in the sea as stable nitrates. How was it that nitrogen continued to make up 78 percent of the air around us? In short, the composition of Earth's atmosphere is wholly unexpected, indeed utterly "improbable" in Lovelock's word, but the planet has somehow maintained its surprising balance of gases over long periods of time.

The mystery was how Earth could be a stable planet when it is made up of such unstable parts. If methane and oxygen persist, it must be that something keeps replacing these gases in the atmosphere as fast as they are being destroyed in chemical reactions. The answer came to Lovelock in a flash of insight. Living organisms must be providing the constant flow of gases that regulated the atmosphere. The Earth owed the dynamic balance of its atmosphere, its stable instability, to a collaboration of all life in concert with inanimate geological and chemical processes. Indeed, in creating and constantly maintaining itself through this grand metabolism, the planet as a whole exhibited behavior, according to the definition of influential biological theorists, fundamental to life. The Earth is thus not merely a planet with life but in the deepest sense a living planet.

Some four decades later, scientists may still debate the extent to which Earth does or does not resemble something living, but the once radical notion that "life is a necessary and active player" in a single, interconnected, self-regulating system is now unquestioned and provides the foundation for scientific investigation of the Earth system. And in its continuing search for life on planets beyond our solar system,

NASA today values chemical analysis of the atmosphere as a primary way to recognize its presence.

The encounter between this awesome Earth metabolism and an increasingly gargantuan human economy that dominates the planet has given rise to the unique challenges of the planetary era. We still do not have a complete understanding of why we are facing this escalating planetary emergency. The answer is not simply that we have altered nature with fossil fuels and carbon dioxide. Living things inevitably change their environment in the process of living, as Lovelock recognized, by taking in resources from the surrounding environment, transforming them, and returning the waste. But with the advent of the modern industrial economy, the human enterprise over the past two centuries has been transforming the Earth on a *scale* and at a *speed* that is mind-boggling and unprecedented. This great burst of profound and still accelerating change has been altering everything everywhere on Earth. By its sheer magnitude, human activity is transforming the oceans and the composition of the Earth's atmosphere and unhinging this grand global metabolism that maintains conditions necessary for life. Based on available evidence, it appears that the Earth system has never experienced change of these types on this scale and at such rates before. This is how we emerged as a planetary force—by "pushing the Earth system well outside of its normal operating range." When I contemplate the possible implications of this radical experiment, I often see in my mind's eye a fish in an aquarium fiddling with the controls that regulate the tank. If "the unifying force of an age is its predicaments," this radical experiment with the planetary metabolism is our predicament, the unifying force of our planetary era.

Scientists may describe the nature of this radical experiment with the Earth's metabolism, but they cannot explain how and why we ended up in this predicament. The answer to these questions lies in unique historical developments in modern Western civilization over the past two centuries. In the broadest terms, the predicament of this age stems from the astonishing growth in the scale of the human enterprise that began in the middle of the eighteenth century with the Industrial Revolution in England. This expansion has been propelled not only by machine power and a new mode of production, the factory system, but by the imperatives of capitalism and competitive nationalism, sudden access to great stores of energy in the form of fossil fuels and the use of new kinds of raw materials, the advances in science and technology, and the modern Western ethos of progress, economic growth, and control of nature. This dynamic of industrial capitalism was indeed revolutionary, for it did not merely bring an acceleration of economic growth, but, as the historian Eric Hobsbawm observes, a new *process of growth* characteristic of the modern era: "self-sustained economic growth by means of perpetual technological revolution and social transformation." As the society transformed itself to foster this process and entrepreneurs plowed income back into new factories, additional machines, and innovations that improved and expanded production, growth became self-perpetuating and, in the words of economist W. W. Rostow, "more or less automatic." The momentum built on successive waves of new technologies, new communication and transportation networks, new kinds of economic organization, and an unabating maelstrom of social upheaval as "human society became an accessory of the economic system." All these elements reinforced each other, creating the power and drive of a unique, dynamic, and evolving cultural phenomenon—modern industrial civilization.

How did this develop into a crisis? Some have attributed our planetary-scale disruption to the rapid growth of the human population, which does indeed play a role in the mounting stress. But the real story over the past two centuries has been the explosive expansion in the world economy—an economic big bang that has made extreme demands on natural systems. The size of the economy has expanded roughly *ten times faster* than human numbers. Moreover, the relationship between population alone and planetary stress is hardly straightforward. Take, for example, the carbon dioxide emissions driving climate change. The world's population is now approaching 7 billion, but a mere 500 million people are responsible for *half* of the carbon dioxide added to the atmosphere each year. Princeton University energy and climate specialist Stephen Pacala states simply that "the climate problem is a problem of the spectacularly rich." By this he does not mean well-off Americans in SUVs, though these contribute their share, but rather the global elite who indulge in private jets and own so many homes around the world that they sometimes lose count.

Over the long course of human history, population and economic growth have tended to rise and fall in tandem, in part because the economy—the goods and services a society produced—relied largely on the power of human muscles. At the start of the nineteenth century, as the Industrial Revolution was gathering steam—both figuratively and physically—people still provided 70 percent of the power even in Europe, a region that enjoyed the luxury of greater numbers of draft animals than other parts of the world. Animals accounted for another 15 percent or so, water and wind for roughly 12 percent, and the new technology of steam engines for the final few percent. The steam engine was the first new tool since windmills for turning available "inanimate" energy—energy not residing in human or animal muscle—into useful mechanical power that humans can employ to do work. The innovation of windmills had come on the scene almost a millennium earlier: One of the earliest reports of this technology comes from a Muslim traveler in the tenth century who saw a windmill lifting water to irrigate a garden in what is now eastern Iran. A windmill captures the kinetic energy of moving air and converts it to useful power; the new steam engines created power by exploiting chemical energy stored in coal and liberated by combustion.

Humans had been using coal for thousands of years before the invention of the steam engine in the eighteenth century, but only on a small scale. In Roman Britain, those assigned to the northern frontier along Hadrian's Wall heated homes, villas, and baths with coal found in that area. In the sixteenth century, when England faced a timber crisis and soaring firewood and charcoal prices in cities, it became the first country to switch from wood to coal. This early turn to a fossil fuel as a replacement for wood provided the solution to a pressing energy crisis, but it did not lead inevitably or directly to the fossil-fueled civilization that would take shape two centuries later.

Two things would have to come together to allow humans to break free of the constraints that had governed previous human societies— the Earth's geological legacy of fossil fuels, which contained vast stores of concentrated energy, and the new tool that provided access to this unimaginably rich energy vault. Before the steam engine, coal could provide heat for homes and industry, but only with the new technology could the wealth of energy be translated into the mechanical power to remake the Earth, both intentionally and unwittingly, and to produce unprecedented economic wealth. The steam engine was the key that first unlocked that power.

In the type of self-reinforcing cycle that would characterize the modern era, the coal-fired steam engine helped make more coal available to build and run more steam engines that could help recover even more coal. British miners had to dig deeper and deeper underground as the demand for coal increased, and in doing so, they quickly hit water, which had to be pumped out before it was possible to dig out the coal. At first, the muscle power of humans and horses powered the pumps, but in 1712, a small coal-fired, steam-powered pump invented by Thomas Newcomen went into operation at a mine in the West Midlands. This new device could pump a remarkable 150 gallons a minute from a 160-foot-deep mine shaft, and, geologist Richard Cowen observes, "At one stroke it turned an enormous amount of British coal (and coal around the world) from reserves to available resources."

Around 1820, the long-coupled trajectories of economic growth and population growth parted ways. This unprecedented divergence occurred as steam engines, rather than muscle, wind, water, or animals, powered the second phase of the Industrial Revolution and the process of industrialization grew to dominate the economic organization of the societies where it had taken root. Propelled by access to fossil energy, the world economy began to expand far faster than human numbers and would accelerate over time to attain astonishing exponential growth. The barest statistics here are simply breathtaking. In the past two centuries, while the human population increased more than *sixfold* from 1 billion to now more than 6 billion, energy use has escalated more than *eightyfold*, and the world's economy (measured in 1990 international dollars) has grown roughly *sixty-eight-fold*. It took all of human history for the global economy to reach the 1950 level of over $5 trillion; in this decade, the world economy expanded that much in a single year.

In the beginning, the great hope had been that this industrial leap and economic growth would usher in an era of plenty and consign poverty to history, but despite the unimaginable wealth generated, the modern era has fallen far short of this promise. The world economy now produces eleven times more per person than it did in 1820, yet 77 percent of the world's people remain poor. In recent decades, the benefits of economic growth have flowed increasingly to the rich—rich people, rich corporations, rich countries—and incomes and the standard of living have steadily risen for many people who live in industrial countries. At the same time, however, the divide between rich and poor is growing within countries and between countries—a trend that not only is appalling in itself but also is a threat to social stability and world peace. The world today is one where the rich fifth of humanity spends $2 on a cappuccino while half of the people in the world struggle to survive on that amount per day. These top 20 percent, who live in the United States, Europe, Japan, and other developed countries, command

85 percent of the world's income and consume, on average, twice as much grain and fish, three times more meat, nine times more paper, and eleven times more gasoline than those living in developing countries. A generation ago, this top 20 percent were thirty times richer than the bottom 20 percent; today they are more than seventy-eight times richer. In 1950, the world had two poor people for every rich one; today the ratio is four poor for every rich person. In this world of gross disparity, the world's three richest people have more money than the combined GDP of the forty-seven poorest countries. Despite robust growth in countries like China and India, inequality overall has grown with economic globalization. Economists such as Nobel laureate Paul Krugman may tout the benefits of this economic growth and boast that "the human race has never had it so good," but this explosive economic growth and growing inequality have increased strains in human societies as well as in planetary systems. The record of the industrial era is at best mixed, and the balance of its costs as opposed to its benefits is indeed debatable.

The speed of this economic big bang has been as important as its magnitude. Around the time I was born, at the end of World War II, the human enterprise moved into fast forward, accelerating in an explosive expansion that has yet to abate. Of all the astonishing facts I've encountered, none compare with this: As modern industrial civilization churned across the face of the Earth in the second half of the twentieth century, it transformed the planet as much *in the span of a single lifetime* as did changes wrought by five hundred generations of our forebears through the ten millennia that saw the beginning of settled life, the rise of agriculture, and the advent of complex civilization. Imagine: half of the human transformation of Earth in fifty short years. Half.

If history had a g-force, we would all be pinned immobile against our seats like space travelers moving at warp speed. And like the adventurers in *Star Trek* who shoot to the speed of light and disappear into another dimension, the past fifty years in exponential overdrive have propelled modern civilization beyond past historical experience into the realm of the utterly unprecedented. Indeed, the familiar graphs of historical and environmental trends over time—carbon dioxide emissions, affluence, energy consumption, water use, paper consumption, the number of automobiles, economic growth, fertilizer and water use, ozone depletion—all trace a path that climbs gently upward from around 1800, and then in the mid twentieth century, the line suddenly shifts into vertical liftoff like a rocket. To emphasize this profound acceleration of human impact on all fronts since 1950, the authors of a definitive volume on global change presented twenty-four such graphs in a memorable two-page centerfold, charting how and why this half century has been "unique in the entire history of human existence on earth."

I first encountered this stunning statistic about the change in my own lifetime almost two decades ago, yet my mind still reels when I pause to contemplate what it says about this time on Earth. It is almost impossible to grasp the magnitude and speed. The churning historical whirlwind that roars around us is all that most people alive today have ever known. Its dynamic of exponential growth and change continues to press toward the fast-approaching physical limits of our ultimately finite planet, and it lies at the heart of the conundrum of the human future.

The question remains: What explains the explosive growth that catapulted humans into the planetary league? One frequent answer is human ingenuity and the technological and institutional innovation that propels self-sustained economic growth. However, those who consider not just human economic systems, but the requirements of complex systems in general, arrive at a different explanation. Like all living systems, human societies maintain themselves by the continuous flow of energy, and the more complex the society, the more energy per capita it takes to maintain it. Indeed, as the pioneering chemist, Nobel laureate, and maverick critic of mainstream economic theory Frederick Soddy urged over eighty years ago, "The flow of energy should be the primary concern of economics." In the time since, other economic thinkers, such as Nicholas Georgescu-Roegen and Kenneth Boulding, have further illuminated the relationship between the human economic system and the physical principles of thermodynamics and energy, but prominent economists still fail to give due weight to the physical basis of wealth.

In the industrial era, economic growth and rising energy consumption have gone hand in hand. Over time, it is true, modern industry has become more efficient in its use of energy and materials, so we get more economic growth for the amount of energy invested. But since the economy keeps growing at a rate greater than its gains in efficiency, the total demand for energy and the pressure on natural systems increase inexorably. The United States, for example, now uses 40 percent *less* energy than it did two decades ago to produce the same amount of goods and services, but the nation's energy use has nevertheless grown by 27 percent. By growing 3 percent a year, the economy has managed to outpace efficiency gains of 2 percent a year. The story of ever greater efficiency is, therefore, a subplot in the larger story of exponential growth fueled by an extravagant and ever growing expenditure of energy. Without this rising consumption of fuels and electricity, concludes energy historian Vaclav Smil, none of the technical and institutional innovations "would have made much difference." The abundance of cheap, concentrated energy available in fossil fuels put power behind this ingenuity and fueled the ongoing innovation. The switch to coal, oil, and gas brought a millionfold increase in the fuel available and allowed the engines of the modern era to revolutionize life in a way that would not have been possible had they been fueled by wind, wood, or peat. More than anything else, what transformed the world in the twentieth century was "enormous energy flows."

Without this torrent of energy, for example, the world's population could never have expanded fourfold in the twentieth century. The amount of land devoted to agriculture grew by only one third during this period, yet the global harvest multiplied sixfold, mainly because of a breathtaking *eightyfold* increase in the use of energy to produce food. New crop varieties that fueled the Green Revolution required more synthetic fertilizer, pesticides, and irrigation to deliver their impressive yields, and all of this demanded far more energy. "Industrial man no longer eats potatoes made from solar energy," as the celebrated ecologist Howard T. Odum put it. "Now he eats potatoes partly made of oil." Today the U.S. food system uses ten kilocalories of fossil energy to deliver a single kilocalorie of food energy to the supermarket. American farmers who produce this food expend three kilocalories of fossil energy for every kilocalorie of harvest.

Fully half of the energy invested in such industrialized agriculture has gone to produce artificial fertilizer through the Haber-Bosch process, which captures nitrogen from air. Though this technology is not widely appreciated today, it was arguably the most important invention of the twentieth century, earning Fritz Haber and Carl Bosch Nobel Prizes for chemistry in 1918 and 1931. Haber became known as the wizard "who created bread from air." Between 1939 and the end of the century, the worldwide use of synthetic nitrogen fertilizer jumped from 3 million tons to more than 85 million tons—a twenty-eight-fold increase—and without the additional food that artificial fertilizer made possible, 40 percent of the current human population simply wouldn't be here. This process today generally uses natural gas; large-scale commercial alternatives that do not depend on fossil fuels are not available.

The industrial era has been in every way a "flamboyant period." Tapping into underground deposits of fossil fuels was like coming into a vast inheritance—the stored sunlight from countless summers over thousands of millennia distilled from ancient forests, swamps, and marine plankton into glossy ribbons of coal, high-pressure pockets of flammable natural gas, and viscous subterranean lakes of black oil. For humans, who had long survived on a modest solar wage gained from agricultural crops, woodlands, water, and wind, cracking into the vaults of fossil fuels gave sudden access to extravagant resources. The unprecedented power of industrial civilization has come from a phenomenal conflagration almost beyond all imagining as this legacy of ancient swamps, forests, and sea life has gone up in flames. All around us, the great fire of the industrial era burns—inside car engines and jets streaking across the sky, in power plants and factories, in lumbering bulldozers and soaring construction cranes, in lawn mowers, snowmobiles, and Jet Skis, in chainsaws chewing through forests around the world.

From this perspective, the past two centuries have been something of a blowout party, a binge that has been immensely liberating and exhilarating but that, there is ample reason to believe, can't go on forever. Not only is this fossil legacy finite, but we've been spending it at an astonishing and accelerating rate. In the 150 years since the first oil wells, the modern appetite for this convenient, concentrated fuel has already exhausted at least 40 percent, perhaps approaching half, of the Earth's known oil reserves. Even with maximum efficiency, our fossil-fueled civilization may not even last through this century. In the long view, "All of this is just an interlude." Moreover, its environmental burdens—evident not only in rising carbon dioxide levels but also in the turmoil of exponential change—will likely bring it to an end far sooner. The great fire has created the best of times in human history, at least for some, and it may yet create the worst for all.

The dangers arising from this explosive high-energy growth over the past two centuries fall into two general categories, which might be described as *slow death* and *surprises*. The slow-death threats are the familiar problems of vanishing species, eroding land, deteriorating soil, water depletion, loss of forests, pervasive contamination of food webs, and the cumulative burden of human activities on natural systems. Humans began to alter the environment on a large scale long before agriculture and complex societies emerged, first with the use of fire. Through countless millennia, hunting groups in many parts of the world periodically set fires to drive game animals over cliffs, into rivers, or onto peninsulas where it would be easier to kill them. They also burned to hold back forests, to open travel corridors, to make it easier to harvest chestnuts in Tennessee, hazelnuts and olives in Europe, and acorns in California. Our ancestors burned routinely to encourage the growth of grasses that supported favored types of game animals, a practice that created open woodlands and grasslands in many parts of the world. With the rise of agriculture, the domestication of plants and animals, the conversion of large tracts from forest to fields, and the rise of complex civilizations, our impact grew further, but the effects were generally limited to the surrounding area. With the explosive growth of human numbers and industrial civilization over the past two hundred years, humans have come to dominate the Earth to such an extent that some scientists have dubbed this time, in geological fashion, the Anthropocene.

Earth at Night, a composite picture taken by NASA from space, says as much as a chapter of statistics about the pervasive human presence. Etched by our lights, the outlines of continents and islands stand out against the absolute black of the oceans. One can see the world's cities spattered across the land like brilliant sequins and trace the Siberian railroad by the city lights along its route. Here are just a few measures of the human burden: We have transformed half of the Earth's surface. We claim half of all the accessible freshwater. We are taking so many fish out of those oceans that half of the fish stocks around the world are being stretched to the limit, while many of the rest have already collapsed or are in danger of collapse. Severe degradation is claiming 24.7 million acres of land suitable for farming every year. Since World War II, roughly 43 percent of the vegetated area on Earth—12.3 billion acres—has been lost to soil depletion, desertification, and the destruction of tropical rainforests. These growing pressures undermine the renewal powers of ecosystems and erode the foundations of human life, but the scale of their impact has been largely local and regional—until recently. The Global Footprint Network calculates that humans were using half of the planet's renewable capacity in 1961. A half century later, human demands have grown to the equivalent of 1.3 planets, which means we are exhausting natural systems faster than they can regenerate.

It is only in the past decade that scientists have recognized the magnitude of the second danger, the surprises—unpredictable, abrupt changes that now threaten as the human enterprise has attained a planetary scale. Modern industrial civilization is now deeply engaged with complex global systems, radically altering mechanisms on which life has depended for hundreds of millions of years. Climate change is without question the most widely recognized problem arising from disruption of the Earth system, but the cycling of nitrogen, phosphorous, and sulfur is suffering from changes as profound as those affecting the carbon cycle. Human activity—which converts unreactive nitrogen in the atmosphere into its reactive, biologically available form—has been adding more nitrogen to terrestrial ecosystems than the amount contributed by natural processes, pushing the global nitrogen cycle "far from its preindustrial steady state." Humans have similarly disrupted the natural phosphorous cycle by mining phosphate deposits for use in fertilizers and adding this nutrient to natural systems at five times normal rates. The burning of fossil fuels is adding sulfur to the Earth system at three times the natural rate.

This overload of sulfur and reactive nitrogen has already caused one surprise that emerged in the late 1970s, the phenomenon of acid rain, a devastating problem that occurs hundreds of miles downwind from the coal-fired power plants and other pollution sources causing it. This corrosive assault carried in rain, snow, and fog has damaged historic buildings, bridges, and other human structures, destroyed fish and other aquatic life in lakes, and led to the death of trees in forests around the world. Over the past two decades, acid rain has diminished in some regions as the U.S., Canadian, European, and Scandinavian governments imposed measures to control pollution. Elsewhere, however, as developing countries have been rapidly acquiring cars and building new coal-burning power plants to meet the growing demand for electricity, the problem is severe and growing in scale. In China, one third of the country's land now suffers from acid rain, which is degrading the soil and threatening farmers' livelihoods. Surprises are, by definition, unexpected, but their likelihood increases rapidly as human activity pushes fundamental cycles far beyond their normal range. And because the carbon, nitrogen, and sulfur cycles are linked in the great Earth metabolism, the consequences of overload in one of these nutrients may well show up in unexpected ways.

Scientists understand less about the possible dangers to human well-being of nitrogen than of carbon dioxide. Just as rising carbon dioxide levels are leading to global warming, this great excess of nitrogen from human activity is causing "global fertilization," upsetting the nutrient balance in ecosystems and creating havoc. Humans create nitrogen overload in natural systems by manufacturing artificial fertilizer, by planting legumes that can take nitrogen from the air and "fix" it into a biologically accessible form, and by burning fossil fuels that release nitrogen gases. The rapid growth in the use of man-made fertilizers in the past sixty years has more than doubled the amount of nitrogen flowing down rivers to the sea, causing an explosive growth of algae that ultimately results in "dead zones," where oxygen levels are too low to sustain animal life. When these algae die, the tiny plants provide food for bacteria, which use up much of the oxygen in the surrounding water in the process of decomposing them. While some fish may be able to flee as oxygen levels plummet, bottom-dwelling creatures, such as eels, crabs, skate, and flounder, can't escape. In Mobile Bay in Alabama, the bottom dwellers, trying to outrun advancing areas of low-oxygen water, end up stranded at the water's edge, gasping for life. The Gulf of Mexico's annual Dead Zone may be the most notorious, but summer brings dead zones to Chesapeake Bay on the East Coast, to the coastal areas of Washington and Oregon, and to some forty other places in U.S. coastal waters. As fertilizer runoff and other nutrients have poured into coastal waters, each passing decade has seen the dead zones around the world double in number. Today the number is approaching 150, and dead zones have been spreading over increasingly larger areas.

As this overload of biologically active nitrogen cascades through natural systems, it begets many other significant problems as well: It plays a role in smog and the formation of health-threatening fine particles. It generates the greenhouse gas nitrous oxide, which, molecule for molecule, is three hundred times more effective in trapping heat than carbon dioxide and lasts in the atmosphere for 120 years, and thereby contributes to global warming. And nitrous oxide also aids in the destruction of the protective stratospheric ozone layer.

Despite the current obscurity of the problem, fertilizing the Earth may prove as problematic as warming it, for like climate change, it can trigger "alarming and sometimes irreversible effects." As the world struggles to feed a growing human population in coming decades and the use of fertilizer continues to grow, managing the damaging effects may pose as great a challenge as managing emissions from fossil fuels. Because of our dependence on fixed nitrogen to grow food, and because of the absence of substitutes, nitrogen promises to be the next urgent global problem, one even more intractable than climate change.

Industrial civilization continues to push us further and faster into the realm of the unprecedented. As it has shifted in and out of ice ages through four glacial cycles over the past 420,000 years, the Earth has demonstrated a strong pattern of self-regulation, which has limited the extent of these climate extremes: The levels of greenhouse gases in the Earth's atmosphere did not exceed roughly 280 parts per million for carbon dioxide and 750 parts per billion for methane. Today the atmospheric carbon dioxide level stands far outside the natural range at 385 parts per million—one third higher than at the start of the Industrial Revolution—and methane levels have more than doubled, to over 1,700 parts per billion. The most recent data from Antarctica show that current carbon dioxide levels are not only *far higher* than at any time in the past 800,000 years, they are climbing *faster than ever before*, a rate one British scientist described as "scary." In the fastest increase in this long ice-core record, the atmospheric carbon dioxide level rose 30 parts per million in roughly one thousand years. In this fossil-fuel era, humans have recently added that much to the atmosphere in seventeen years.

Given the vastness of the system and the great stretches of time, it is no doubt difficult to imagine how this astonishing global change relates to our daily lives, save as a looming threat. All of this, however, is not only about a changing Earth, but also about an altered human prospect. The incremental slow-death changes damage particular places on Earth and may impair the health and possibilities for those living there. The whirlwind of global change, however, threatens the functioning of Earth as a whole. By disrupting not just places, but the planetary *system*, people living in all parts of the world now confront a shared human future. In one way or another, it will matter to everyone if China burns its coal and Americans continue to drive energy-dinosaur SUVs and the spectacularly rich contribute extravagantly to climate change.

The flamboyant period of the human career has not only pushed the Earth system well outside of its normal operating range; it promises to confront us in the coming decades with conditions on Earth that are beyond anything in the 200,000-year evolutionary history of modern humans, or in a worst case, beyond anything encountered by our more distant ancestors over the past 5 million years. Without question, we have now been thrust into a new chapter of the human story. We live in a time like no other.

Lessons from the Ozone Hole

If any single event marked the arrival of the planetary era, it was unquestionably the sudden appearance of a yawning hole in the ozone layer over Antarctica more than twenty years ago, a stunning surprise that confronted humanity with the first life-threatening breakdown of a planetary system. It proved a close call that we survived thanks not to scientific prowess but to mere luck. Although world leaders rallied to "solve" the immediate crisis, the larger lesson about the unique perils of the planetary era was lost. Those who set the course for this civilization proceeded onward, seemingly oblivious to the fact that the world had changed fundamentally.

The ozone layer is one of Earth's wonders. The next time you find yourself looking up at an expansive blue sky, take a moment to consider our utter dependence on things unseen. Above your head, some ten to fifteen miles up in the stratosphere, sunlight and oxygen are engaged in a chemical dance of constant creation and destruction that maintains a thin veil of gas known as the ozone layer. The modern understanding of the ozone layer would have delighted the pre-Socratic philosopher Heraclitus, who saw the world as process, declaring in the sixth century B.C. that "it is in changing that things find repose." This critical part of the great planetary metabolism has endured for some 600 million years through a dynamic balance of opposing processes emerging within a restless flux. "Opposition" does, as Heraclitus believed, bring "concord," and "out of discord comes the fairest harmony."

The ozone layer, which began forming some 2 billion years ago, was a major development in the evolution of the atmosphere and the Earth system, and much in the history of life on Earth since has hinged upon on its unending dance. Without this high-altitude concentration of ozone, which is a rare and poisonous form of oxygen, the Earth would not be the green planet astronauts see from space, nor would our kind of life be among its inhabitants. If it were to vanish, deadly ultraviolet radiation produced by the sun would penetrate the atmosphere and damage the DNA of the plants and animals that inhabit·the Earth's land surfaces, jeopardizing the life we see all around us. The ozone layer, in short, provides essential protection and made life on land possible.

Sometime around 1976, the great global experiment of the modern industrial era began to unhinge this dynamic balance. The cause was a man-made chemical accumulating in the Earth's atmosphere that began to destroy stratospheric ozone at a rapid rate. Within a few years, ozone levels over Antarctica plummeted in spring, and a "hole" suddenly opened up in the ozone layer over the South Pole—a gaping loss that in successive years grew more severe and expanded across an area larger than North America. Since then, the ozone hole has opened and closed in an annual rhythm, healing to some extent by midsummer, only to open again when sun returns to the South Pole with the coming of each austral spring in September. In winters that are cold enough in the Arctic, an ozone hole now develops over the North Pole as well, though the loss is typically shorter and less severe than that over Antarctica. Beyond the poles, the ozone layer has been suffering from chronic erosion ranging from 3 to 6 percent from the level of overhead ozone in 1980.

Amid the whirlwind of global change over the past six decades, the Antarctic ozone hole signaled the giant leap humans had taken on Earth and our new status as a planetary force. We had stepped across a fateful threshold into an unprecedented relationship with the Earth and into a new historical era. If there had been any question, the ozone hole left no doubt that pushing the Earth system "outside of normal operating range" could have dramatic consequences.

In 1930, Thomas Midgley Jr., a brilliant, quirky American inventor and industrial pioneer, had a solution to an urgent problem: dangerous refrigerators. In the early decades of the twentieth century, the move from wooden iceboxes cooled with blocks of ice delivered by the iceman to electrically powered, chemically cooled metal refrigerators had run into a serious obstacle. The gases used to cool the refrigerators—ammonia, methyl chloride, and sulfur dioxide—were all poisonous; refrigerator leaks were killing people. Chicago and surrounding Cook County, Illinois, saw twenty-nine cases of methyl chloride poisoning from faulty refrigerators within a single year in the late 1920s. Growing concern about the hazards prompted health authorities in some cities to require warning labels on refrigerators.

Midgley, who worked for the auto industry's legendary engineer and inventor Charles Kettering at General Motors Research Corporation, had been enlisted in the search for a nonpoisonous refrigerant for Frigidaire, a refrigerator manufacturer that GM had acquired a decade earlier. Though Midgley wasn't a chemist by training, in a matter of two months of chemical tinkering, he had made a breakthrough. On the final day of the year in 1928, Frigidaire patented Midgley's formula for a fluorine-based compound—chlorofluorocarbon, or CFC-12—and DuPont was soon at work figuring out how to manufacture the new man-made chemical that would be marketed with similar CFCs under the name Freon.

By the time of the annual meeting of the American Chemical Society in April 1930, Midgley was ready to present his breakthrough invention to the world. He made his case for the safety of CFCs through a theatrical demonstration very much like those that would later make for a popular *Mr. Wizard* science program in the early days of television. At the end of his talk, he set a burning candle and a dish containing CFCs with a flourish on the table in front of him. As the compressed CFCs began to boil off at normal pressure into a white cloud, Midgley leaned low and inhaled the vapors and held his breath. Then he picked up a rubber tube, put it to his lips, and expelled the gas over the candle. The flame died. As everyone could see with their own eyes, CFCs were neither poisonous nor flammable. Midgley reported that inhaling the gas did produce "a kind of intoxication," but it was nothing like the exhilaration of alcohol, for "these fumes do not arouse a desire to sing or to recite poetry." Rather it was a sensation of "deadening."

Six years earlier, Midgley had performed a similar demonstration to defend the safety of another of his inventions—tetraethyl lead, a gasoline additive that was developed to solve the problem of "knocking," which caused noise, jerkiness, and loss of power in internal combustion engines. As Standard Oil of New Jersey set to work on developing a method for manufacturing this additive, workers at its pilot plant began to exhibit symptoms of severe mental illness, including panic, delirium, and paranoia. One threw himself out a window; another became so violent and deranged that he had to be removed from his home in a straightjacket. More than thirty workers at the facility in Elizabeth, New Jersey, were put under observation because of what the newspapers dubbed "loony gas," and at least four eventually died, as autopsies revealed, from tetraethyl lead poisoning. With the first death, the story of this "odd gas" that makes people insane made the front page of the *New York Times*. Soon health specialists were asking questions about danger posed by the lead that would be coming out of auto tailpipes. As alarm grew, Midgley, the inventor of the controversial product, was summoned to help defuse the crisis. Although Midgley himself had been poisoned by lead in the course of his research and had taken a lengthy vacation the previous year to allow his lungs to recover, he willingly vouched for the safety of his lead additive. At a news conference at Standard Oil's New York City headquarters, he complained that it was difficult to get workers to follow proper procedures for handling lead, suggesting they were to blame for their illness. Then he produced a bottle of tetraethyl lead, and, to prove the compound was harmless, the *New York Herald Tribune* reporter wrote, he "washed his hands with it and held the bottle to his nostrils for more than a minute. He insisted the fumes could have no such effect as was observed in the victims if inhaled for only a short time." Despite well-documented concerns about the dangers posed by lead, the additive went into gasoline and, as a consequence, generations of children around the world would suffer neurological damage.

Unlike tetraethyl lead, the new refrigerants, CFCs, appeared to be in fact inert and as harmless as Midgley claimed to workers, consumers, and the environment. Midgley's inventions earned him prestigious awards and fame, as well as the presidency of the American Chemical Society in the 1940s, but in time, their impacts on human health and the ozone layer would come to make him infamous. He ranks as a uniquely influential figure, as environmental historian J. R. McNeill observed: He "had more impact on the atmosphere than any other single organism in earth history."

Scientists did not yet fully understand the ozone layer when Midgley was synthesizing CFCs. Half a century earlier, those studying the wavelengths of sunlight reaching Earth observed the absence of some parts of the spectrum, notably the shorter waves of ultraviolet light, or UVB, so they began searching for the answer to the mystery of the missing ultraviolet. In 1879, a French researcher theorized that some gas in the atmosphere was absorbing UV. A year later, W. N. Hartley, a chemistry professor at the Royal College of Science in Dublin, concluded that this gas was ozone and noted, moreover, that the atmosphere above the Earth must contain greater concentrations of ozone, because there wasn't enough at ground level to account for the absent UVB. By the 1920s, French physicist Charles Fabry had arrived at a decent estimate on the location of the "ozone layer" in the stratosphere and the amount of ozone it contained. Then in 1927, Oxford University physicist and meteorologist Gordon Dobson developed a device—the Dobson ozone spectrometer—that could measure the ozone overhead, an instrument that decades later would allow British scientists to detect the hole in the ozone layer. Dobson began his measurements at Oxford and later initiated a global monitoring network that helped demonstrate that the ozone layer enveloped the entire planet.

About the time Midgley was demonstrating both his showmanship and the apparent safety of his "miracle" refrigerant, a British geophysicist on the other side of the Atlantic was putting forward the first explanation of how the ozone layer forms and is constantly being regenerated high above. Sydney Chapman's historic paper "A Theory of Upper Atmospheric Ozone," published in the *Memoirs of the Royal Meteorological Society* in 1930, explained how ultraviolet light acts on oxygen molecules—produced at the Earth's surface by photosynthesizing plants—that had percolated upward into the stratosphere. As Chapman correctly theorized, short, high-energy ultraviolet light rays, invisible to the human eye, interact with oxygen molecules (O_2) and split them into two separate oxygen atoms. Each solo atom then joins with a whole oxygen molecule to form a molecule of ozone, the rare three-atom form of the gas. Ozone, however, is itself unstable and vulnerable to destruction by ultraviolet light, so it soon splits back into a two-atom oxygen and a single atom of oxygen, which tend to react quickly with available partners and create ozone again. In the opposing process, the ozone layer loses ozone when a lone oxygen atom encounters ozone and they recombine to form two ordinary oxygen molecules. So high overhead in a chemical contra dance, an endless whirl of oxygen molecules breaking apart and coming together has maintained the ozone layer in a dynamic but steady state for hundreds of millions of years. This process absorbs the damaging UVB rays that would otherwise reach the Earth's surface and converts them into heat—a diversion that warms the stratosphere and at the same time protects life.

Ozone is exceedingly rare overall in the Earth's atmosphere, and most of it, some 90 percent, occurs in the stratosphere. The remaining 10 percent forms near the ground from the interaction of sunlight with nitrogen oxide pollutants emitted by power plants and auto tailpipes and the fumes of such volatile organic compounds as solvents and gasoline. Since ozone is poisonous even in small concentrations, ground-level ozone in the air we breathe, often called bad ozone, poses a serious hazard during episodes of ozone smog in the hot, sunny summer months. It causes throat irritation, chest pain, and difficulty breathing, and those who live with repeated exposure in such notorious smog hot spots as Los Angeles can suffer permanent scarring of lung tissue. Whether ozone protects or harms depends, as real estate agents are reputed to say, on "location, location, location." What seems most remarkable about the role ozone plays in the Earth system is how much depends on so little. The ozone that helps sustain life exists at a few parts per million generally and at only 10 parts per million in the thick of the ozone layer.

One might get the impression from Chapman's description of the ozone layer that "good ozone" is a creation of sunlight and high-altitude chemistry alone, but its appearance above the Earth is as much about evolution as about chemistry. The advent of the ozone layer is part of a larger story of how the atmosphere came to be what it is today—a history that scientists have been piecing together over the past three decades into a revolutionary new understanding. Although researchers continue investigating and debating exactly when and how events have unfolded, it is possible to sketch a provisional outline.

This epic origin story is fascinating in its own right, for it reveals the awesome creativity of Earthly life and the ways it has managed to endure over the long run. But this investigation bears on our current dilemma, as well. The emerging history of the Earth provides important insights into its nature and the behavior of the planetary metabolism humans are pushing beyond normal operating range. Surprising discoveries have been upending long-held assumptions about the stability of the Earth system and the limits of global change.

This ongoing exploration of Earth's past has also greatly expanded the common understanding of evolution, showing that life's progress has entailed a great deal more than the origin of plant and animal species that Charles Darwin's theory sought to explain. The atmosphere itself has evolved over the past 4.6 billion years of Earth's existence, and life has played a leading role in the process. Living things *created* the current balance of gases and in doing so provided conditions that allowed an ozone layer to form and opened an evolutionary door leading to larger, visible forms of life, such as our own.

Earth history's overall has been dominated by long periods of stability punctuated by a series of major transitions. The largest and most momentous of these, from a human perspective, was the shift beginning some 2.4 billion years ago in which free oxygen in the Earth's atmosphere rose from virtual nonexistence to levels sufficient to support animal life. The action of living things triggered this transition, inaugurating the era of "the living atmosphere," and as Lovelock recognized, Earth's life continues to maintain the atmosphere today.

The immense sweep of time entailed in Earth's history easily overwhelms a mind that measures time by the span of a year, a decade, or a human lifetime. Placing Earth's history within the framework of a single year is a useful way to get a fix on what happened at what point within this unimaginably long story. Thus, the planet would take shape on January 1, and the first life would emerge on March 4, or some 3.8 billion years ago. By 2.2 billion years ago, July 9, photosynthesizing cyanobacteria, which release oxygen, are permanently shifting the balance in the Earth's atmosphere.

Reconstructing the story of this evolution is a work in progress, subject to constant revision as scientists find new evidence of the Earth's deep past and seek to explain how it fits into the larger story. Recently, for example, the prevailing view that oxygen rose to the current level of 21 percent in a single surge and has remained steady for the past 600 million years has come under challenge. Now, various lines of evidence, including new methods of estimating oxygen levels in the distant past and an influential model developed by Yale geochemist Robert Berner, are making a persuasive case that oxygen levels have varied greatly during this period. Based on a comparison of this oxygen history and the pattern of evolution and extinction, some scientists now believe that oxygen has been a major driver in the evolutionary process, especially during the "age of animals," which began 542 million years ago when complex, multicellular life surged onto the scene.

The Earth's oxygen-rich atmosphere had its origin in an ancient crisis that arose as the planet's expanding life kept running into constraints in its struggle to find the means of survival in a world of finite resources. The account of how early microbes overcame a series of limits entails three interrelated stories of evolution: first, the innovation by microbes themselves to acquire food and gain access to increased energy; second, the global impact of these microbial innovations and the evolution of the atmosphere; and finally, the evolution of complementary kinds of microbes that together form the Earth's grand cycles—a critical step that made life an enduring feature of Earth rather than a short-lived experiment that arrived at a dead end when essential resources ran out.

In the earliest days after life emerged, it is possible that simple bacteria enjoyed a free lunch, living directly on biochemicals created by the unique conditions of the period and available in the surrounding environment. Over time, however, as their multiplying numbers depleted this chemical bonanza, some microbes developed processes for converting sugars found in the Earth into the principal form of energy that fuels organic life—adenosine triphosphate, or ATP. One of the first of these was the fermentation metabolism, still employed today by the yeast that makes bread rise and by the bacteria essential for making wine and beer. Other metabolic strategies emerged early on as well, including microbes that fix inert atmospheric nitrogen into organic nitrogen compounds needed to make proteins and microbes that convert sulfate into hydrogen sulfide to obtain energy.

Under increasing pressure to find new sources of food, some bacteria eventually hit upon a whole new way to make a living—the revolutionary breakthrough of photosynthesis, which freed them from reliance on resources found in their immediate environment. Through photosynthesis, they could *produce* food—the carbohydrate sugar—from the ubiquitous ingredients of sunlight and air. This step, in the judgment of the eminent microbiologist Lynn Margulis, was "the most important single metabolic innovation in the history of life on the planet." Up to this point, microbes had survived by consuming and breaking down available food, so this new ability to produce food from readily available carbon dioxide and hydrogen gases was something of a microbial industrial revolution. This lineage of early photosynthesizers survives today in green and purple bacteria inhabiting oxygen-poor salt flats or mudflats.

These first photosynthesizing microbes flourished until another limit began to loom—a hydrogen crisis. Readily available hydrogen had become increasingly scarce, in part because of the great demand of the now vast communities of microbes and also because of a decline in atmospheric levels as the Earth lost some of this light gas to space. In the midst of this growing crisis, one kind of microbe, cyanobacteria, suddenly found a way to tap the virtually limitless supplies of hydrogen locked up in water, the H in the familiar chemical formula H_2O. In a leap that would redirect the trajectory of life on Earth, this evolutionary innovator developed a mutation that made it possible to use the energy from captured sunlight to split water molecules, freeing the hydrogen from the oxygen. These microbes then used the hydrogen for photosynthesis and dumped the leftover oxygen, which is a very toxic gas, into the surrounding environment.

Some scientists are now convinced that microbes evolved photosynthesis far sooner than previously thought. Over the past decade, researchers investigating some of the oldest rocks on Earth have been challenging and radically revising the accepted account of conditions on the young Earth and the dates for critical steps in the evolution of life. In the emerging view, the early Earth was far more hospitable than previously thought, making it already suitable for life as soon as 200 million years after it formed—4.4 billion years ago, or January 15 in our evolutionary year. Moreover, Danish scientists studying ancient sea floor sediments contained in ancient rock from Isua, Greenland, have made the bold claim that microbes not unlike present-day cyanobacteria were already carrying out this more complex, oxygen-producing method of photosynthesis as early as 3.7 billion years ago, March 12. If this is the case, anoxygenic photosynthesis, which does not produce oxygen and is thought to have evolved first, must have emerged at an even earlier date.

Access to the vast quantities of hydrogen through water-based photosynthesis fueled an explosive expansion: cyanobacteria spread to every corner of the Earth that offered water, sunlight, and carbon dioxide. The oceans must have been swarming with these microbe multitudes, which had expanded from local communities into a dominant global presence and, as their oxygen pollution began to accumulate in the atmosphere around 2.3 billion years ago, into a planetary force.

The global pollution crisis that ensued triggered a series of events that turned the world upside down. Life had originated and expanded in a world that was virtually free of oxygen, so as oxygen levels rose, the atmosphere grew deadly. Much of existing life was swept away in a global catastrophe, according to Margulis—an "oxygen holocaust" that was "by far the greatest pollution crisis the earth has ever endured." Those that managed to escape retreated to anaerobic refuges— airless recesses beneath mud and soil where they have carried on down to the present. By 2 billion years ago, July 26 on our imagined calendar, the balance of gases in the Earth's atmosphere had shifted permanently.

Scientists have not fully explained this great transition to the oxygen-rich atmosphere we know today, and they are still hotly debating various theories about how this pulse of free oxygen figured in the tumultuous planetary events that followed. The span between 2.3 billion and 580 million years ago—from the beginning of July to mid-October—was a perilous period in Earth's history, marked by wild instability in which conditions swung back and forth between extremes of cold and heat. Three, perhaps four, extreme ice ages with temperatures averaging 58 degrees below zero F alternated with interludes of equally extreme hothouse conditions with average temperatures of 122 degrees F. One leading theorist, California Institute of Technology geobiologist Joe Kirschvink, believes that this period of climate extremes began with a runaway glaciation that locked the entire planet in solid ice and almost brought the great evolutionary story to an abrupt end. Whether these ice ages were so extreme that they caused the oceans to freeze solid, as this "snowball earth" theory contends, is still under debate, but Kirschvink and others now believe that it was the great evolutionary leap of the cyanobacteria and their oxygen pollution that triggered this first planetary deep freeze some 2.3 billion years ago.

This climatic emergency likely involved not only oxygen but also methane, which was a vital component in the early atmosphere. One of the puzzles in the history of the Earth had been why average surface temperatures were comparable to those of today even though the sun 3.5 billion years ago was far cooler. If the sun today were to dim and give off only 75 percent of its current light, the oceans would freeze solid, but they did not freeze under such conditions on the early Earth. The answer to the puzzle, some scientists now believe, is the abundance of methane in the atmosphere, produced largely by the anaerobic bacteria that flourished before oxygen levels rose and forced them into retreat. Methane, the gas that powers kitchen ranges and still oozes today out of swamps, rice paddies, and landfills, is a far more potent greenhouse gas than carbon dioxide, at least seventy-two times stronger over a period of twenty years, so it helped maintain the Earth's temperature in the era of the "weak sun."

According to this theory, the oxygen released as the wildly successful cyanobacteria spread across the face of the Earth began to destroy the methane greenhouse effect that maintained the Earth's temperature. First, the increase in oxygen would have killed off large numbers of anaerobic microbes that produce methane. Second, oxygen would destroy methane in the atmosphere by reacting with it to produce carbon dioxide and water. By one estimate, oxygen levels only a fraction of 1 percent of today's levels would have been enough to overwhelm the methane greenhouse. The increasing presence of oxygen may have destroyed the methane in perhaps as short a time as 100,000 years, plunging the Earth's temperature to 58 degrees below zero F.

Long before the "snowball earth" theory emerged in the early 1990s, the poet Robert Frost published a short verse in 1920 debating whether the world will end in fire or ice. Frost decided he holds with those who favor fire, yet allows "for destruction ice / Is also great / And would suffice." And indeed, ice might have sufficed to ring down the curtain on life. What may have finally come to the rescue, in a planetary version of *The Perils of Pauline*, was the Earth's volcanoes. Despite a deep sheet of ice, scientists theorize that they continued to erupt and spew out carbon dioxide, a greenhouse gas that helps warm the Earth. Since the microbes and rocky land surfaces that serve in normal conditions to remove carbon dioxide from the atmosphere were icebound, scientists believe this volcanic carbon dioxide had a chance to build up in the atmosphere, restore a sufficient greenhouse effect, and eventually warm the Earth once again.

This harrowing narrative raises provocative and unsettling questions. Did an evolutionary "advance," water-based photosynthesis, trigger one of Earth's worst climate disasters and nearly bring an end to life on Earth? Is innovation, which can prove life-enhancing or lethal, unavoidably a game of Russian roulette?

In one of the great ironies in Earth history, the very process of photosynthesis that may have put all of life in jeopardy by creating climate chaos also helped ensure that life would be more than a fleeting phenomenon. With the spectacular success and worldwide expansion of cyanobacteria using water for photosynthesis, a new shortage loomed. They might soon exhaust the other essential ingredient for food production—the carbon dioxide in the atmosphere. Then, inadvertently, the cyanobacteria creating this shortage helped spur another innovation that would solve the problem—the recycling of the carbon dioxide. The food production of photosynthesis was an initial step in the formation of one of Earth's grand cycles, and this recycling of a critical ingredient was the second major step.

As the cyanobacteria flooded the environment with the oxygen waste from their water-based method of photosynthesis, other ever-inventive microbes responded to rising levels of oxygen by evolving a way to use it in a new kind of metabolism—respiration—which took in oxygen to help turn sugar into ATP and released carbon dioxide as waste. Today a wide variety of species, including humans, require oxygen, but the first oxygen-breathing *aerobic* microbes first had to develop defense mechanisms—not unlike a snake's evolved defenses against its own venom—before life could exploit the new opportunities offered by oxygen. The immediate benefit for these oxygen-breathing microbes was a huge energy bonanza. While a fermenting microbe like yeast can produce only two molecules of ATP energy for every molecule of sugar consumed, an aerobic microbe that breathes oxygen can obtain as many as thirty-six molecules of ATP from the same sugar molecule—a stunning eighteenfold increase.

The leap in available energy provided by respiration would open a whole new frontier of evolutionary opportunity, the possibility of creatures larger and more complex than the one-cell wonders that had innovated, prospered, and transformed the Earth. Equally important, respiration and other processes that recycle essential ingredients for life allowed the microbes to survive long enough to evolve into these visible creatures. Switching from one resource to another as they depleted available supplies, the bacteria would have eventually exhausted the supply of resources and arrived at a dead end. But with the evolution of respiration, life, which had been proceeding along a one-way street to eventual extinction, closed the circle. The recycling of carbon dioxide through respiration made it possible for industrious photosynthesizing cyanobacteria to make food for eons without ever running out of resources, just as the waste oxygen from photosynthesis continuously replenishes the oxygen required for respiration. If this recycling process were to stop today, plants and other photosynthesizing life could exhaust all the available carbon dioxide in the Earth atmosphere in roughly a decade.

Establishing these grand cycles was one of the great problems early life had to solve in order to have a future. "Life, as we know it," explains evolutionary theorist Marcello Barbieri, "requires a community of complementary organisms," for all of the grand planetary cycles that together keep the wheel of Earthly life in spin involve several steps that *no individual organism can accomplish alone.* Enduring life is inescapably a joint enterprise and, as the scientist and essayist Lewis Thomas put it, "The urge to form partnerships, to link up in collaborative arrangements, is perhaps the oldest, strongest, and most fundamental force in Nature."

Much of this story seems to bear on the dilemma of modern industrial civilization, as well. At the moment, it is a linear metabolism on a one-way street, consuming extravantly and discarding vast quantities of waste. We are far from closing the circle. As Barbieri observes, lots of organisms have appeared in the history of life, "but only a few managed to form natural cycles. In the long run, only those survived."

The oxygen crisis made way for our kind of life in a variety of ways. High energy demand and complexity go hand in hand. In creating an oxygen-rich atmosphere and evolving respiration, the microbes gained access to high energy flows and whole new realms of possibility. "Without oxygen," one researcher observed, "the most sophisticated life on Earth would have been green microbial scum." As the cyanobacteria poured out oxygen, some of it began floating up to the stratosphere, where it interacted with sunlight and formed the ozone layer—a development that made possible the evolution of land-dwelling species. Moreover, the climatic instability that coincided with the rising oxygen and the wild swings between ice ages and hothouse conditions—"freeze-fry events"—may have pushed evolution into overdrive, leading to the explosion of complex, visible forms of life between 575 and 525 million years ago—a momentous event in the history of life on Earth known as the Cambrian explosion. At the end of October on the calendar, the "age of animals" had arrived. Around November 29, 400 million years ago, life, now shielded from deadly UVB radiation by the ozone layer, moved from the oceans onto land. Then, in the waning minutes of this evolutionary year—December 31 at roughly 11:50—modern humans finally joined this ancient commonwealth of life.

Long-standing assumptions have fallen in the ongoing investigation of Earth's past. Although individual species have come and gone throughout this story, so that more than 99 percent of species that have existed are now extinct, Earthly life overall is far more robust than once thought. The climate system, however, is apparently far less stable. If perturbed, the Earth system is capable of extreme change, and sometimes, as the Antarctic ozone hole made clear, all it takes is a small push.

James Lovelock's mind was always puzzling about something. Around the same time in the mid-1960s that the British scientist and inventor was consulting to NASA about detecting life on Mars, he began to wonder about the dense summer haze that swallowed the blue sky and obscured the pleasant vistas afforded by the English countryside where he lived. While England had been notorious for smoky pollution in winter caused by coal burning, Lovelock remembered clear skies in summer during his childhood—in fact, right up until 1950. Sometime afterward, this pall gradually descended, making Lovelock wonder about the nature of the "new miasma." Through observations on the density of the haze, he determined the summer murk in rural Wiltshire in southern England was almost as bad as the legendary smog in urban Los Angeles. That kind of photochemical smog—a mixture of ground-level ozone and other harmful pollutants formed from the interaction of sunlight, vehicle exhaust, and volatile industrial chemicals—had first been described just a decade earlier. And another decade would pass before air-quality specialists recognized that pollution often traveled long distances from its source and could do damage even in remote places.

Convinced that the haze was indeed smog, Lovelock next sought to figure out where the air that brought smog was coming from. He suspected an urban industrial area, but how to establish that unequiv-ocally? The best way to solve the mystery, he decided, would be to an-alyze the air on clear and hazy days, looking for some chemical that would have virtually no sources outside of cities. If this chemical was missing on clear days but present in the times when the sky was ob-scured, it would show that the haze in the countryside was blowing in from urban areas. The compound that he thought filled the bill was one of the CFCs. Moreover, since he had invented the electron cap-ture detector, he was an old hand with the device that could measure the slightest concentrations of CFC in the atmosphere, down to 1 part per trillion.

The commercial use of Midgley's CFCs had followed quickly on the heels of their invention. In a $7 million advertising campaign in 1931, Frigidaire proclaimed the wondrous features of its new and improved refrigerator cooled by a "miracle" refrigerant that ranked as "one of the most outstanding scientific achievements of our times." By 1935, other cooling chemicals had virtually vanished from use; most of the 8 million refrigerators sold that year in the United States, by Frigidaire or its competitors, used Freon, CFC-12.

Even before the Freon-cooled refrigerator hit the market, Midg-ley had a meeting with the inventor of air-conditioning, Willis Carrier, to discuss how CFCs might advance this industry, which had likewise been frustrated by the lack of a safe, efficient coolant. Carrier's inven-tion was already cooling factories, theaters, and office buildings, but he had not found a way to make small air conditioners for home use. Car-rier returned from the meeting with a sample of one of Midgley's other compounds, CFC-11, which would revolutionize the world of air-conditioning. By 1932, Carrier had a CFC-cooled, self-contained home air conditioner on the market—the Atmospheric Cabinet. In the decades that followed, the technology would not only transform daily life but also make way for the massive population shift in the United States toward the hotter southern climes of the Sunbelt.

During World War II, CFCs found an entirely new use as the gas in pressurized cans that could blast pesticides into the air as fine, penetrating aerosol droplets. In the summer of 1941, as it seemed increasingly likely that the United States would enter the fight, army medical officials were already preparing to combat what had been the most formidable enemy in wartime—insect-borne diseases, such as malaria and typhoid, which generally claimed far more victims on the battlefront than bombs, guns, and artillery. After a public presentation in July 1941 demonstrating a new aerosol can using CFC-12, the army set out to develop portable aerosol insecticide sprays that troops could use for protection on the front lines—a mission that took on immediate urgency as thousands of U.S. troops began to succumb to malaria as the Pacific campaign got under way in August 1942. At its worst, the disease was killing five victims for every one brought down by Japanese fire. Westinghouse, which had shifted its production from refrigerators to military supplies, such as torpedoes, found a way to convert a container developed for injecting Freon into refrigerators into the Bug Bomb, a three- by five-inch spray can filled with pyrethrum, a botanical insecticide derived from chrysanthemums, and Freon 12. The army ordered emergency production and within three months began delivering the insect sprays to troops serving in the South Pacific. Later in the war, the army added another powerful insecticide, DDT, to the aerosol insecticides supplied to soldiers, making it such a potent weapon that one army medical officer dubbed it "the atomic bomb of the insect world." U.S. companies would produce some 40 million Freon-propelled Bug Bombs in the war effort.

As news stories about the Bug Bomb emphasized the importance of Freon to this powerful weapon against malaria, Midgley was again lionized for his invention of a "miracle" and a "wonder" chemical. A writer for the *New York Times* Sunday magazine in 1945 rhapsodized about the virtues of Freon and its lengthy chemical name: "A double delight is dichloro-difluoromethane with its thirteen consonants and ten vowels," for it "brings death to disease-carrying insects and provides cool comfort to man" in the baking months of July and August.

Once the war ended, the Bug Bomb made its way into peacetime commerce with breathtaking speed. As the formal surrender of the Japanese was taking place in September 1945, Bug Bombs filled with pyrethrum, DDT, and Freon were already hitting the shelves at Gimbels department store in New York City and proving a runaway success. The first 2,500 sold out in two hours, and the consumer stampede was on—not only for Freon-propelled DDT, sold with such names as Real-Kill, but for any other products that could be delivered by means of a convenient aerosol spray can. Once the aerosol industry solved initial problems like leaky, clogging valves, America launched into the aerosol age of air fresheners, hair spray, and Reddi Whip in lightweight aluminum cans propelled by a combination of Midgley's Freons, CFC-11 and CFC-12. As air-conditioning and aerosol cans became as much necessities for Americans as refrigerators, CFC production in the postwar years shot upward, growing from some 100,000 tons a year in the late 1950s to 1.5 million tons by 1986. With each *psssttt* of a spray can, CFCs floated off into the atmosphere.

Armed with his electron capture detector, James Lovelock pursued the mystery of the summer haze in 1969, taking measurements of haze density, wind direction, and levels of CFC-11 in air samples in southern England and later on the southwest coast of Ireland facing the Atlantic. His investigation first established that he was correct in his suspicion that the haze blanketing rural areas was man-made and later confirmed it was indeed smog containing "bad ozone." Polluted air traveling from Europe brought ozone levels exceeding the U.S. Environmental Protection Agency's safe limits even to remote western Ireland.

But the solution to one mystery brought another. Lovelock's electron capture detector was finding CFC-11 in samples of clean Atlantic air—50 parts per trillion. Perhaps the answer was simply that the winds had carried the gas across from the United States, but then again, there was a far more interesting possibility. Could inert CFCs be building up in the Earth's atmosphere? Lovelock was off again like a beagle following another scent caught by his electron capture detector in a whiff of ocean air.

By the autumn of 1971, Lovelock was aboard the research vessel *Shackleton* and headed to the southern oceans. After an initial rejection, he had managed to secure a berth and a minimal amount of funding from a government agency to measure chemicals in the atmosphere, including dimethyl sulfide and CFC-11, during the passage from England to Montevideo, Uruguay, on the eastern coast of South America. Near the equator, the ship crossed the low-pressure zone—known to sailors as the doldrums and to scientists as the Intertropical Convergence Zone—that encircles the Earth and acts as a barrier to the mixing of air between the hemispheres. In the morning, when Lovelock looked out at the clouds and the sky, everything had a stunning clarity; the industrial haze that enshrouds the Northern Hemisphere had vanished. His next samples confirmed that Southern Hemisphere air was indeed cleaner. CFC levels, which had been measuring 70 parts per trillion en route, had dropped to 40 parts per trillion. But finding such levels south of the equator meant that CFCs were *everywhere* in the Earth's atmosphere and "accumulating ineluctably." Lovelock published his findings in a paper in the leading scientific journal *Nature* and discussed the possible implications. He would later come to regret his flat declaration that CFCs posed "no conceivable hazard."

The notion that humans might possibly do significant damage to some vital, invisible gas overhead in the stratosphere was already in the air in 1971, but the concern had no relationship to Lovelock's work measuring CFCs. If ordinary people had heard about the ozone layer at all, it was during the highly polarized battle over the SST—the supersonic jet that would travel in the stratosphere—proposed in the early 1960s by an Anglo-French partnership and later by the U.S. government. The U.S. opposition, which gathered steam after 1967, centered initially on the path of shattering sonic booms the planes would create as they traveled faster than the speed of sound. The campaign to stop the SST featured prominently at rallies on the first Earth Day in April 1970, and such well-known figures as aviation pioneer Charles Lindbergh questioned the wisdom of the project.

As the fight went on, scientists began to raise new concerns about the effect of the plane's exhaust on the stratosphere. The Europeans would go on to build the Concorde, but Congress ultimately voted to halt government funding of the U.S. SST program in March 1971, more because it did not make economic sense than because it threatened the integrity of the ozone layer. In prompting further study of the effects of supersonic aircraft and other human activities on the atmosphere, however, the controversy left a lasting legacy. Although atmospheric science was still in its infancy, Dutch-born scientist Paul Crutzen had already warned that the ozone layer was vulnerable not only to the exhaust of supersonic aircraft in the stratosphere but also to substances originating far below.

Lovelock's report about CFCs accumulating in the atmosphere prompted mild curiosity among scientists and a degree of anxiety among CFC producers like DuPont. The company put up money to finance new studies on CFCs, and DuPont officials gathered industry scientists, academics, and Lovelock at a school in central New Hampshire in November 1972 to examine "the ecology of CFCs." The discussion, however, focused primarily on CFCs' possible toxic hazard to humans in normal household use and confirmed the long-standing conviction about their overall safety. The question of whether the growing abundance of CFCs might have large-scale consequences, Lovelock later recalled, was touched upon only briefly in his own talk. He noted that CFCs would act as a potent greenhouse gas if concentrations were to rise eventually from parts per trillion into the parts per billion range. Because that would be a tenfold increase from the 1972 levels, it seemed a long-range concern. Meanwhile, the scientific studies funded by DuPont on the ecological effects of CFCs proceeded and likewise concluded in 1974 that the CFCs accumulating in the lower atmosphere posed no hazard.

Word of Lovelock's findings began spreading in atmospheric-science circles long before the publication of his results in 1971. F. Sherwood "Sherry" Rowland, a professor of chemistry at the University of California at Irvine, was attending a workshop on atmospheric chemistry in January 1972 when he first heard about the measurements, as well as about Lovelock's suggestion that CFCs might prove useful to researchers as a trace gas for tracking the movement of air in the atmosphere. If ordinary cleansing processes, such as rain or degradation in sunlight, do not remove CFCs from the lower atmosphere, what, Rowland wondered, would eventually happen to them? Based on some experience in photochemistry—the interaction of molecules and light—one thing seemed clear: It would be wrong to assume they would remain inert forever. Nothing, however, suggested to Rowland that CFCs would pose any danger; the concentrations were, after all, so small. Their ultimate fate was simply an intriguing scientific question.

Some twenty-one months later, Rowland suggested the CFC question as one of several possible projects for his new postdoctoral student, Mario Molina, a native of Mexico whose prominent family had sent him as a youth to European schools to foster his early ambition to become a research scientist. Like Rowland, Molina immediately recognized that CFCs were not likely to remain intact if they reached the stratosphere, where ultraviolet light would break the molecules apart. "At the very least," Molina would later say, "I thought it was very bad manners just to release these chemicals without knowing what would happen." He elected to tackle the CFC project, a decision that would set Molina and his adviser off on a path that eventually led to a Nobel Prize.

Step by step, Molina traced the fate of CFCs and answered the questions they had asked at the outset. They established that CFCs did not break down in the lower stratosphere, as Lovelock's measurements had suggested. They calculated that these sluggish compounds could drift around for decades—between 40 and 150 years—before filtering up into the upper stratosphere, where they would encounter ultraviolet light and break down. This photochemical reaction would yield a free chlorine atom. By December 1973, the work seemed ready for publication, but Rowland and Molina decided to address one final, relatively easy question. What happens to the free chlorine?

Determining the answer was a matter of straightforward gas-phase chemistry, but as Molina worked out the interactions between the chlorine atoms and ozone, he discovered something utterly surprising. The atoms of chlorine would destroy ozone by taking one of the three oxygen atoms that make up ozone, but that was only half the story. The chlorine atoms themselves would not be destroyed in the process, because the ozone layer has many individual oxygen atoms floating around. These single oxygen atoms would break the joined chlorine and oxygen atoms apart, thereby freeing the chlorine atom to destroy more ozone. In this way, one chlorine atom could destroy ozone molecules over and over again in a powerful catalytic chain reaction. The free chlorine would remain in the stratosphere for months, and each atom would keep chewing through the ozone layer the whole time. By Molina's calculations, a single chlorine atom could knock out 100,000 ozone molecules. This was interesting, but, given the scale of global processes, Molina assumed the effect would be "negligible" until he brought industry production figures on CFCs into his calculation. Rowland and Molina were floored by the answer and kept checking and rechecking to make sure it was correct. The destructive efficiency of these chlorine atoms plus the yearly release of 1 million tons of CFCs added up to a "very significant global environmental problem." Once they made their way to the stratosphere, CFCs would wreak havoc.

Publishing their findings in a seminal article that appeared in *Nature* on June 28, 1974, Rowland and Molina warned that the ozone layer at midlatitudes would suffer increasing erosion in the decades ahead if CFC production continued to increase as it had been, at the rate of 10 percent a year. This would result in a 5 to 7 percent loss by 1995 and as much as 50 percent by 2050.

The *Nature* paper set off a debate about the reality and magnitude of the CFC threat and what, if anything, should be done—a controversy that waxed and waned over the course of the decade that followed. Rowland and Molina's theory prompted a flurry of research and several scientific assessments, but instead of settling the question, the new efforts tended to add to the uncertainty. The more scientists investigated the chemistry of the atmosphere, the more complex the matter of ozone depletion became. In the midst of this, the U.S. government moved in 1977 to phase out nonessential uses of CFCs by the end of 1978, which was, in effect, a ban on their use in aerosol cans. Meanwhile, as one historian of the ozone battle noted, "Estimates of ozone depletion fluctuated wildly from almost 20 percent to no loss at all." The National Research Council of the National Academy of Sciences issued four reports during this period, which affirmed the fundamental validity of Rowland and Molina's theory but offered different estimates regarding the seriousness of the threat. In the last of this series of assessments, issued in 1984, the NRC panel concluded that ozone was *less* vulnerable to CFCs than previously thought. The threat had been overestimated. The eventual loss of ozone would be limited to 2 to 4 percent. One science magazine summed up the report in a headline: "Ozone: The Crisis That Wasn't."

The ozone hole blindsided atmospheric scientists. After more than a decade of debate and intense research, there was a sense that scientists had a basic grasp of how the atmosphere worked and could make reasonable assessments about possible dangers. That something this dramatic should emerge shortly after a reassuring assessment by a leading scientific body left little doubt that our ignorance about the complex systems remained greater than our knowledge.

The arresting news that 40 percent of the ozone layer had disappeared in spring over Antarctica arrived in May 1985 in a paper written by Joseph Farman, Brian Gardiner, and Jonathan Shanklin of the British Antarctic Survey. The study, published in *Nature*, reported measurements of ozone directly overhead taken with a spectrometer, the instrument invented by Gordon Dobson, as well as levels of CFC-11 and CFC-12, which had risen markedly since the early 1970s. The authors plotted the ozone and CFC data together in a single graph that shows matching curves: As CFC levels rise, ozone levels plummet. The British team had been extremely cautious about going to publication with their findings because none of the published data from NASA's Nimbus 7 satellite, which was monitoring ozone levels, indicated anything out of the ordinary.

The reaction among those who had been studying the ozone threat was stunned disbelief. None of them had ever heard of Joe Farman or the British Antarctic Survey. And how could this data *possibly* be correct when NASA had been monitoring ozone since 1978 with far more sophisticated instruments?

The notion that such a gaping hole in the ozone layer could appear in a matter of weeks was inconceivable in 1985. The mere suggestion that this was possible, recalled Susan Solomon, an atmospheric scientist with the U.S. National Oceanic and Atmospheric Administration (NOAA) and a leading investigator of the ozone hole, would have been deemed "preposterous and alarmist." Up to this point, the strenuous search had detected little if any damage to the ozone layer. Moreover, a huge ozone loss over Antarctica after the darkness of winter was about the last thing anyone would anticipate.

When NASA scientists reviewed their satellite data, they discovered that their two ozone-monitoring devices had indeed recorded the dramatic ozone loss that Farman had reported. However, the NASA computer used to process the 140,000 satellite readings taken daily and covering the whole surface of the Earth had not been including extremely low readings in the compilation of data for analysis. The scientists had programmed the computer to include only those readings within what they assumed to be a realistic range. The computer treated readings outside this range as anomalies, which it stored and flagged. Other American researchers taking ground measurements in Antarctica had not recorded such low ozone concentrations, reinforcing the prevailing assumption that low readings were likely wrong. NASA scientists assumed at first that the low satellite readings were instrument errors. Many have analyzed the reasons why one of the world's leading scientific institutions, using the most sophisticated measuring devices available, had failed to detect the hole. A large part of the answer is surely that the ozone loss exceeded anything considered possible.

Once the satellite data confirmed the dramatic loss over Antarctica, scientists began a frantic search for the cause, making risky high-altitude flights into the ozone hole to measure the chemistry of the stratosphere. The measurements left little doubt that the culprit was Midgley's CFCs and that the rapid destruction of ozone resulted from runaway catalytic chemical reactions involving rare stratospheric clouds that form in the extreme cold and dark of winter over the South Pole. This process, involving heterogeneous chemical reactions on ice particles, was, however, not the gas-phase ozone-depletion chemistry described in Rowland and Molina's historic paper. It was far more destructive.

Spurred by the sudden appearance of the ozone hole, in September 1987, representatives from 183 nations joined in the Montreal Protocol, an international treaty mandating the phaseout of chlorofluorocarbons, halons, methyl chloroform, methyl bromide, carbon tetrachloride, and other ozone-depleting substances. The concentration of these chemicals reached a peak in the Earth's troposphere in 1995 and is now declining in this lower level and in the stratosphere as a result of the treaty and its amendments. Nevertheless, in the spring of 2006, Antarctica experienced the largest and most severe ozone hole on record. All the ozone in the ozone layer between 8 and 13 miles above the South Pole was "virtually gone," NASA scientists reported. Scientists attributed the new lows to record cold conditions as well as to very high levels of ozone-depleting chemicals. The recovery of the ozone layer over Antarctica will take decades and, according to a recent estimate, will not be completed until about 2065.

Even worse, it appears that there may be new twists to the ozone saga still to come. The Montreal Protocol relied in large part on a technological fix, the substitution of a new generation of refrigerant chemicals, hydrochlorofluorocarbons or HCFCs, that posed less threat to the ozone layer but were unfortunately powerful greenhouse gases that could drive global warming. This has suddenly become a significant concern as increasing prosperity in China and India has led to booming sales of air conditioners and much greater production of these problematic substitutes. In the face of mounting international pressure, the 191 nations that are now parties to the Montreal Protocol agreed in September 2007 to accelerate the phaseout of HCFCs by a decade. On another front, scientists fear that ozone depletion, perhaps by cooling the stratosphere, may be contributing to a shift in wind patterns around Antarctica that could have global consequences. With shifting winds, warmer waters are now flowing beneath ice shelves, causing melting and "surprisingly rapid changes" to a piece of the Antarctic ice sheet that is roughly the size of Texas and holds enough water to raise sea levels about twenty feet.

This historic ozone treaty has been heralded as an environmental success story and a model for grappling with other planetary problems, including global warming. That is, of course, true. But in some respects, the appearance of the ozone hole and its aftermath is also a story of purblindness, a failure that has left us ill prepared to grasp the perils of climate change and other escalating threats to planetary systems. Political leaders did not appear to understand the larger significance of this alarming hole in the sky or to recognize that the human enterprise had crossed into uncharted territory. There was little inclination to push beyond the level of physical symptoms and confront profound and pressing questions about how humanity had arrived at this juncture. Was something inherently problematic in modern industrial civilization's now global-scale enterprise? They mustered a technological fix— different, but not trouble-free chemicals—and hurried on with business as usual.

The modern era arrived at a dead end with the ozone hole, because this fateful encounter with planetary systems flatly contradicted the modern understanding of how the world works. Since the Scientific Revolution, Western culture has pursued knowledge and power on the assumption that they would give humans ever greater control over a nature that was assumed to be a passive resource to be exploited and directed with impunity. In 1930, the year that refrigerators cooled by CFCs first appeared on the market, the president of the American Association for the Advancement of Science, Nobel laureate Robert A. Millikan, assured the public that any fears they might harbor about scientific progress were unfounded. In an essay titled "Alleged Sins of

Science," published in *Scribner's Magazine*, he declared, "One may sleep in peace with the consciousness that the Creator has put some foolproof elements into his handiwork, and that man is powerless to do it any titanic damage." Just a few years before the ozone hole emerged, Robert L. Sinsheimer, a prominent molecular biologist and commentator on matters of scientific risk, noted that our scientific and technological civilization has proceeded *on faith* in the resilience, even the benevolence, of nature—"the faith that nature does not set booby traps for unsuspecting species." Writing at the time of the ozone debate in the late 1970s, Sinsheimer discussed CFCs and the ozone layer, assuming that it would require an "extended, large-scale release of fluorocarbons" to cause massive depletion.

We could hardly have been more wrong. The ozone hole was not only a surprise; it far surpassed any worst-case scenario that scientists had thought plausible. Although the destruction occurred in the stratosphere, as Rowland and Molina had theorized eleven years earlier, it proceeded through a completely different chemical process. Most important, this hole in the sky was not caused by gross pollution but rather by vanishingly small concentrations of man-made chemicals in the Earth's atmosphere. The quantity of total chlorine from CFCs and other compounds that caused this loss of the ozone layer over an area larger than North America can be measured not in parts per million, but parts per billion. Lovelock had measured parts per trillion on the *Shackleton*, but as the production of aerosol cans and the emissions of CFCs increased in the 1970s, the levels in the atmosphere rose past 1 part per billion. When the total concentration of atmospheric chlorine hit 2 parts per billion in the mid-1970s, the destruction of the ozone layer commenced over Antarctica. Parts per trillion and parts per billion are minute quantities. One part per billion is the distance of one foot on a trip to the moon or one second in the span of thirty-two years. Two parts per billion is a quantity barely detectable until Lovelock invented his electron capture detector in 1957. In the wildly nonlinear world of complex systems, a tiny insult caused swift destruction on a vast scale.

The human enterprise survived this first encounter with planetary systems thanks only to dumb luck, argues Paul Crutzen, who shared the Nobel chemistry prize with Rowland and Molina in 1995 for his pioneering work showing that nitrogen oxides from fertilizers and supersonic aircraft could damage the ozone layer. Had the problematic refrigerants been engineered not with chlorine but with bromine, a similar chemical and possible alternative, the world would have faced catastrophic destruction of ozone everywhere in all seasons and significant harm to land-based forms of life. In his 1995 Nobel acceptance speech, Crutzen explained that, atom for atom, bromine is one hundred times more destructive to ozone than chlorine because it does not require unusual conditions for its activation. The rapid ozone destruction caused by CFCs over Antarctica, by contrast, depends on heterogeneous chemical reactions on the solid or supercooled liquid particles found in rare polar stratospheric clouds, such as those found over the South Pole in the total darkness of winter. "I can only conclude that mankind has been extremely lucky," Crutzen concluded. "It was a close call."

The saga of the ozone hole is a history rife with dangerously mistaken assumptions about the nature of the world. It is a story of how chemicals considered to be among the safest ever invented turned out to be some of the most dangerous. The widely used man-made compounds at the center of this narrative seemed for almost half a century to be an unalloyed miracle of modern technology before proving instead a nightmare of planetary scale. Scientists first investigating possible hazards found none because they looked in the wrong place. Later, after the discovery that the ozone layer might be in jeopardy, leading scientific experts concluded that the danger had been overestimated and that eventual ozone loss would be minimal. This reassuring assessment appeared just months before the inconceivable destruction of ozone over Antarctica came to light. Scientists not only failed to predict the ozone hole, they were slow to see it when it opened up, in part because it was considered beyond the realm of the possible.

This remarkable story illuminates not only the perils of disrupting the planetary metabolism but equally the inability of science to predict the consequences with accuracy. This lapse speaks both to the dangers of preconceptions in unprecedented times and to the limits of knowledge—of irreducible scientific ignorance regarding how these complex, nonlinear systems might respond. Unpredictability arises from the intrinsic nature of these dynamic, complex natural systems, which are open, self-organizing, and indeterminate. It is not a problem that can be remedied by bigger models or more data. All of this underscores the radical uncertainty at the heart of humanity's new relationship with Earth. This initial encounter between the human enterprise and Earth's grand metabolism proved our first significant brush with global disaster. Our planetary debut could well have been our evolutionary exit.

In pushing Earth systems beyond normal operating range, modern civilization has thrust us into a potentially deadly game in which wild cards rule. In the planetary era, we must, therefore, expect the unexpected, consider the unthinkable, and recognize that the future may hold events worse than "the worst-case scenario." The wager in this includes human survival.

Today we are facing climate change without having understood these crucial lessons from our encounter with the ozone hole. Here again, expectations have proved wrong because climate change is advancing harder and faster than anticipated by almost all scientists.

The most urgent lesson to be learned is this: Nature *does* set booby traps. The image of a predictable, imperturbable machine that we inherited from the modern era is giving way to the very different picture of a dynamic and potentially volatile living system. Domination, it turns out, has not given humans dominion. Immense power has not given us control. To understand this is to recognize that the modern era has ended. In the new planetary era, nature will not serve merely as a backdrop for human history but will be a significant and unpredictable player. We should be ready, Crutzen warns, for more surprises.

4

The Return of Nature

There is no reasonable doubt that global warming is under way. The world is changing all around us and on all fronts. Most of us can now see the signs in our own backyards. Here in Massachusetts, the cherry trees bloomed in Boston Garden on New Years Day 2007, and January delivered utterly freakish shirtsleeve weather instead of frozen ponds and ice skating. Reports from the larger world bring pictures of starving polar bears and news about a vanishing ice cap at the North Pole and disintegrating ice sheets in Greenland and Antarctica. These surprisingly swift changes in the realms of ice bode ill for the hundreds of millions of us around the world living on coastal margins, and perhaps for the future of the kind of civilizations humans have developed over the past ten thousand years. What happens at the ends of the Earth now bears directly on the future of my city. If the ice sheets go, which seems increasingly possible, the Freedom Trail, which traces events culminating in the American Revolution—sites of the Boston Massacre, the Boston Tea Party, Paul Revere's Ride—will be lost to rising seas. Since 1900, the Earth has warmed by about 1 degree Fahrenheit, a warming due mostly to human activities. Just one degree has set all this change in motion, and, given the current trends, five times this warming may become inescapable by 2030. In the coming century and beyond, nature, it appears, will become the leading actor in Boston's history.

Though public awareness and concern about climate change

have been growing rapidly, the general grasp of what we might be up against remains vague. Even those who have raised the alarm about the growing danger fail to provide a coherent report of the challenge that lies ahead. Over the past two decades, the most knowledgeable people on the front lines have sought to highlight one aspect or another of the new planetary era, and they have described this crisis in ways that often seem at odds. This seeming confusion is a symptom of our greater predicament, for we find ourselves not only confronting an escalating physical emergency but also trying to get our bearings in the middle of a daunting historical and cultural transition. Are we really "at the planetary controls," as the distinguished environmental leader Gus Speth once put it, emphasizing the extent of human power and dominance? Or are we rather "poking an angry beast with a stick" without knowing how it might "lash out," as the eminent climate scientist Wallace Broecker said, warning of the Earth system's unappreciated power and volatility? Was writer Bill McKibben correct when he claimed twenty years ago that climate change is bringing "the end of Nature"? Can we, indeed, "solve" the problem of climate change, as Al Gore optimistically promises in the Academy Award–winning documentary *An Inconvenient Truth*? The metaphors are telling, for taken together they reflect an evolving understanding and the struggle to find words and ideas to bring this unprecedented time into focus. How much control do we have? And what kind of Nature are we disrupting?

At the heart of this confusion lie two outmoded and dangerous misconceptions: the extent of human power and the character of the nature we inhabit and act upon. If we missed the lesson from the ozone hole, the investigation of ice cores and other records of the Earth's past has propelled an astonishing revolution in scientific thinking about how fast climate can change and also provided added evidence that nature is not as the modern era imagined. The long history of Earth testifies that the modern era has constructed an entire civilization on mistaken assumptions.

It isn't that we fail to recognize that the climate has changed in the past. Everyone knows, of course, that a great ice sheet flowed southward and covered large parts of the Northern Hemisphere in the last ice age and then retreated, leaving the land enriched by its legacy of lakes scoured out of bedrock, kettle-hole ponds formed by blocks of glacial ice, and sandy peninsulas like Cape Cod thrusting out into the sea. But when I was growing up in this glacial landscape of sand banks, moraines, and eskers, we learned that climate changed in the immensely slow, ponderous, deliberate rhythm of geological time. Big change was "glacial," meaning too gradual to be evident during a human lifetime or even ten human lifetimes. Hiking through mountains and woodlands, I reflected on this other kind of time and change whenever I encountered one of the immense granite boulders, some the size of a small house, that are scattered across the New England landscape as if they had fallen from the sky. Geologists call them *glacial erratics*, chunks of rock ripped from distant places, carried by the glaciers southward, and then dropped randomly as the ice melted.

I have a vivid memory of the moment when I learned that this comforting story about slow, deliberate change might be wrong. Wallace "Wally" Broecker, an influential pioneer in the study of past climate change from Columbia University, was speaking at an international conference in Washington, D.C., on the dangers attending human disruption of the Earth's atmosphere, notably destruction of the ozone layer and global warming. This was in 1986 in the wake of the nuclear-freeze movement protesting the insanity of nuclear weapons and of the scientific debate about whether an exchange of nuclear bombs would fill the atmosphere with smoke, destroy most of the ozone layer, and send temperatures downward into a "nuclear winter." Against this backdrop, some at this meeting warned that human changes to the atmosphere could bring on a different kind of slow-motion catastrophe, a "chemical summer."

Broecker's presentation, however, added a whole new dimension to these warnings. I scribbled furiously as he challenged model simulations done on computers that suggested increasing carbon dioxide levels would result in a gradual warming over a century or so. The natural record stored in glacial ice, ocean sediment, and the muck in bogs testified to a different kind of change, which Broecker and others investigating past climate had been piecing together. They now had clear evidence that shifts in the Earth's climate since the height of the last ice age some twenty thousand years ago had been sudden, not gradual. As Broecker continued, it became clear that he really meant *sudden*—sudden in terms of a human lifetime, not sudden according to the conventional scale of measure, geological time. Possibly as suddenly as *within a decade*.

How had this happened? Broecker then thought the clues pointed to the North Atlantic conveyor belt, part of the ocean circulation system that carries tropical heat northward on the Gulf Stream and warms northern climes. This great flow of water and heat, driven by the slide of cold, heavy, salty water from the ocean's surface to the ocean floor near Iceland, had shut down, plunging some regions back into near-ice-age conditions in a matter of years. This was the most astonishing piece of news I had ever heard, but a journalist working on a deadline has no time to reflect. I had to keep writing, and as I did, I had the odd sensation that the floor was giving way beneath my feet as some of my deepest assumptions about the nature of the world around me collapsed. Was it really true that the Earth is wild in ways almost beyond imagining?

Our ability to weather the century ahead will hinge significantly on the way climate change happens.

The common, still widespread notion among ordinary people is that global warming will proceed like an escalator that will convey the snowbelt into an era of palm trees and winter at the beach, perhaps bringing bouts of extreme weather, such as floods, droughts, and hurricanes, as well. It doesn't sound half bad when the TV meteorologist explains in a two-minute tutorial that Boston by 2100 will have a climate like Virginia and be perhaps ten degrees hotter in the summer and fall. But the persistent assumption of progressive, more or less continuous change is also implicit in the mandated cuts in greenhouse-gas emissions set by the 1997 Kyoto climate treaty and in the work of many scientists and policy analysts in the United Nations–sponsored international assessments—the Intergovernmental Panel on Climate Change, or IPCC—about the impacts and potential costs of climate change. Assuming such manageable warming, some economists have long argued that it makes more economic sense to adapt to global warming later, when the global economy will be richer, than to move quickly to cut the emissions driving it now.

The discussion of climate change over the past quarter century generally has been framed in scenarios—exercises conducted on computers that project how rising levels of greenhouse gases will affect temperatures and how ongoing warming will affect other vital things: the number of heat-wave days in summer, the length of the growing season, the rate of evaporation and available water. The typical impact assessment has assumed a limited warming episode in which carbon dioxide levels climb to double the normal level at the time of the Industrial Revolution and then stop—a scenario that will not happen in the real world without a swift and radical shift in the world energy system and human behavior. Such an analysis—typically based on a planetwide average warming of 3.6 to 5.4 degrees F—then determines the effects as the climate system reaches an equilibrium. As carbon dioxide levels rise, temperature also climbs upward in a parallel ascent. The number of sweltering summer days increases, and the growing season lengthens. The rains will become more intense, the droughts more severe, and the tropical storms more powerful. If you consult the graphs showing various possible future scenarios contained in the UN-sponsored international assessments done by the Intergovernmental Panel on

Climate Change, they suggest nothing erratic—no jumps or jagged progression, no sawtooth passages, nothing other than an escalator ascent. Leading climate scientists warn that these projections are "optimistically smooth" and "surprise free." And they all share a larger unwritten and unstated assumption: Despite global warming, the planet will continue to operate largely as it has since the end of the last ice age, and this warming will perturb, but not unhinge, the system.

How realistic are such widespread expectations about the way climate will change? In October 2006, the British government released a report that confronted this question head-on and sought to provide a better-grounded estimate of the stakes in climate change. This analysis—titled "The Economics of Climate Change" and headed by former World Bank chief economist Sir Nicholas Stern—warned that the cost will be much higher than previously thought because most assessment scenarios have not included "the most uncertain but potentially damaging impacts." If the world proceeds with business as usual and temperatures rise 9 degrees F by 2100, the Stern review concluded that up to 10 percent of global output could be lost. In a worst-case scenario with this magnitude of warming, the loss might top 20 percent.

These estimates far exceed previous assessments, and an illuminating section of the Stern report explains why. In a chapter called "Economic Modeling of Climate Change," the authors critique several influential economic models used to generate damage scenarios and detail exactly what is missing from their calculations. The short answer is: far too much. "Most existing studies are . . . limited to a small subset of the most well understood, but least damaging impacts." The Mendelsohn model, done by Yale University economist Robert O. Mendelsohn, assumes, for example, a modest escalator warming, fails to consider the risk of catastrophe, and is extremely optimistic about the ease of adapting to changing climate. *All* the leading economic models have ignored the likely social and economic fallout from climate disruption—the migration of great numbers of people, the risk of military conflict, and the flight of capital investment. This is a serious omission, considering that a development organization estimates that climate change could force 1 billion people to leave their homes by 2050 because of scarce water, failed crops, and coastal flooding. This critique raises deeply troubling questions about the basis for the almost priestlike authority that economists and their models have commanded in policy discussions about climate change. Can assessments that ignore so much of the dangerous uncertainty offer any meaningful guidance to those weighing critical decisions?

The British review team sought to remedy such omissions, adding critical assumptions absent from previous assessments to their computer-run impact assessment model. These include the likelihood that the global average temperature increase by 2100 may *exceed* 2 to 3 degrees C (3.6 to 5.4 degrees F), recent scientific evidence that the natural systems may respond in ways that will speed and amplify the warming caused by greenhouse gases, and "non-market" impacts on human health and the environment. In addition, the authors also claim to incorporate a general estimate reflecting "risk of catastrophe." Like previous economic assessments, however, its conclusions do not fully reflect possible social and economic consequences of climate disruption.

There is, however, something curiously contradictory at the heart of the Stern report. While its text certainly discusses earnestly and at length the possibility of some unspecified "surprise" or "catastrophe," the report nevertheless conveys the overall impression that, even at its worst, climate change will be a manageable, albeit costly, problem. Wally Broecker's "angry beast" is not looming in the background, nor is the possibility of abrupt events that might utterly overwhelm human social and economic systems. Instead, *catastrophe* is defined as the chance of large losses of up to 20 percent of GDP when global temperatures rise 5 degrees C (9 degrees F) above preindustrial levels. The

BBC described the Stern report as "a stark warning," yet its worst-case scenario represents a decidedly optimistic view of how climate change will advance over the coming century. As possible catastrophes go, this is a benign one, indeed. By contrast, Will Steffen, the former executive director of the International Geosphere-Biosphere Program, a research effort that involves ten thousand scientists from eighty countries around the world, judges that modern civilization simply won't be able to adapt to a 5-degree-C warming in a century. Others with a deep knowledge of the emerging science, such as science journalist and author Fred Pearce, worry that its "rosy" view might give the public and policy makers "a false sense of security."

The most vociferous critics of the Stern review, however, have been other economists, who found it insufficiently rosy and protested, among other things, that it exaggerated the economic costs of climate change. One notable exception is Harvard economist Martin L. Weitzman, who states in a provocative paper—"On Modeling and Intepreting the Economics of Catastrophic Climate Change"—that traditional policy advice from economists "should be treated with skepticism." He argues that conventional cost-benefit analysis is "especially and unusually misleading" when applied to climate change, because it ignores "a non-negligible probability of worldwide catastrophe." Weitzman concludes that economists working on climate change "can help most by *not* presenting a cost-benefit estimate . . . as if it is accurate and objective" and by acknowledging instead the "incredible magnitude of the deep structural uncertainties that are involved in climate-change analysis."

One might think that mainstream economists and climate scientists were working on different planets, given the Grand Canyon that divides their views regarding the dangers of disrupting climate. Indeed, this has been one of the many sources of confusion in the public debate. A survey done in the early 1990s of ecologists, atmospheric scientists, conventional economists, and environmental economists found an optimism among conventional economists that appears to border on the irrational. This group judged the global economy to be virtually impervious to severe economic losses, even in the face of extreme climate change and an *11-degree-F increase* in average global average temperatures by 2105. In stark contrast, ecologists and atmospheric scientists estimated that economic losses from extreme climate change would be twenty to thirty times higher. To put this disagreement into a meaningful climatic context, the difference between global average temperatures in the last ice age and today is 9 degrees F. Such a warming would create conditions outside the evolutionary experience of modern humans and the agriculture-based complex civilizations that first arose some six thousand years ago.

This astonishing divide stems not only from differing professional perspectives, but from deeper unspoken assumptions. Economist Herman Daly ran head-on into what is taken for granted among mainstream economists during his time working at the World Bank. Daly, who ranked as the bank's leading heretic for disputing the wisdom of continuing exponential economic growth, was helping prepare a bank report, *Development and the Environment*, when he noticed a puzzling diagram in an early draft. It was titled "The Relationship Between the Economy and the Environment" and consisted of a square labeled "economy" with an "input" arrow entering the box and "output" arrow leaving. Pointing out that there was no "environment" in the diagram, Daly suggested adding a larger box around the economy box to denote the economy's foundation in the larger natural system. This small and seemingly obvious suggestion amounted to a challenge of economic orthodoxy. As Daly recounts his experience, the next draft contained the larger box, but because it was not labeled "environment," it looked like a decorative frame around the box labeled "economy." Daly protested again. The following draft had no diagram at all. When Daly found himself a few months later on a conference panel with then–World Bank chief economist Larry Summers, he asked Summers whether economists should consider the size of the economy relative to the Earth's physical system. Summers, he recounts, dismissed the question outright, declaring, "That's not the right way to look at it."

Paradoxically, mainstream economists who spend their professional lives thinking about material matters seem strangely untethered from physical reality, primarily because of their fundamental assumption that human ingenuity and technology will allow us to transcend anything and everything, even extreme climate change. They would concur with Katharine Hepburn's quintessentially modern declaration to Humphrey Bogart in the 1951 film *African Queen:* "Nature, Mr. Alnutt, is what we are put in this world to rise above." It is in this spirit that the Nobel Prize–winning economist Robert Solow argued that our ingenuity will allow the economy to find endless substitutes for depleted resources so "the world can, in effect, get along without natural resources." It is no doubt in this spirit that prominent economists continue to insist that the Stern report exaggerates the damage that extreme climate change will visit on human systems, such as the global economy, and caution against drastic action.

The mainstream in economic thinking does not take nature seriously. Of course, for much of the twentieth century, leading scientists, relying on the same fundamental assumptions of the modern era, also took it for granted that humans could alter the world with impunity, or at least without catastrophe. The arrival of the planetary era, however, disabused scientists of this modern fiction. Events in the real world shattered the inherited faith in a world without booby traps and spurred urgent investigation of how the Earth system changes. Amid a revolution in our understanding of the world, orthodox economists continue to proffer advice based on demonstrably obsolete assumptions. This priesthood, in its overweening confidence in human power and its underestimation and disregard of nature, has stood firm as a staunch defender of the modern faith. Even the diligent and forthright economic analysts in the Stern review could not bring themselves to imagine genuine catastrophe, a climate spasm disruptive enough to bring down the economic system. Leading climate scientists say this is not only imaginable, but frighteningly possible.

In April 2008, Stern acknowledged that he underestimated the threat of global warming when he compared the worst-case impact to the Great Depression of the 1930s. "The damage risks are bigger than I would have argued," he told the *Financial Times*. "The damage associated with a 5 degree temperature increase would be enormous. We can't be precise about what it would be like, but you can say it would be a transformation." Even in this gloomier view of the risks, Stern does not appear to grasp the full danger of an increase of 5 degrees C in a century. Abrupt climate change and massive human impacts become much more likely if temperatures rise more than 3 degrees C. By the time global average temperatures surge into the range of 5 degrees C, global change authority Will Steffen believes, the Earth system will move to a new state, and modern civilization will collapse.

It has proved extremely hard to let go of this long-standing faith in the generally reliable character of the natural world. For more than two millennia, this has been a deep, powerful, and persistent conviction in Western thought that comes down to us in the Latin dictum *Natura non facit saltum*: Nature does not make a leap. Scholars have traced the roots of this belief to two different sources: the Jewish conception of a lawful nature proceeding according to the dictates of a law-giving God, and the turn in sixth-century-B.C. classical Greece from religious explanations of natural phenomena to naturalistic ones. History has even recorded the date on which the philosopher-scientist Thales of Miletus freed the Greek world from the arbitrary intervention of the gods. On May 28, 585 B.C., as the moon passed between the sun and Earth and cast this prosperous trading city on the coast of Asia Minor into darkness, Thales' contemporaries witnessed the first solar eclipse predicted beforehand. Thus, Thales demonstrated that even dramatic, infrequent, hitherto terrifying events are not supernatural in origin, but lawful in ways that might be discovered. One can explain darkness at midday without invoking a displeased deity on Mt. Olympus.

This idea of lawful gradualness and continuity in nature has since had a long, complex history deeply intertwined with religion, philosophy, and science. The dictum that nature does not make leaps comes down to us most immediately from Charles Darwin in his *Origin of Species*, although it had such prominent earlier adherents in modern thought as the seventeenth-century rationalist philosopher Gottfried Wilhelm Leibniz and the eighteenth-century botanist Carl Linnaeus, father of the modern taxonomic system for classifying living things. Pioneering advocates of evolution, including Darwin's grandfather, Erasmus Darwin, and his French contemporary Jean-Baptiste Lamarck, also shared this conviction of gradualism in the behavior of nature, as did the influential geologist Charles Lyell, who acted as a teacher and mentor to the young Darwin.

The debate between evolutionists and creationists, which still continues today in the United States, started long before Darwin's *Origin* appeared in 1859. Within the framework of the argument in England during Darwin's time, abrupt change often implied supernatural origin. By contrast, things of natural origin emerged in a gradual, orderly, and lawful manner from what came before. Darwin's noteworthy accomplishment was not his advocacy of evolution as a historical fact, but his theory that explained *how* the evolution of new species could occur gradually through "natural selection" without acts of divine creation. In providing a mechanism to explain the diversity of living things, Darwin "completed the revolution in natural philosophy begun two centuries and a half earlier," according to historian and philosopher of science Jerome Ravetz, and was correctly hailed as the Newton of natural sciences. To assume like Darwin that nature did not make a leap reflected a fundamental commitment to rational explanation based on natural laws and implicitly to the modern, material, mechanistic view of the world. In short, to believe the contrary—that nature did make leaps—was *unscientific*.

Now, however, science itself is overturning this stubborn and inherited conviction. Investigation of the Earth's climate history, generated increasingly by concerns about rising levels of carbon dioxide, has shown unequivocally that nature does not proceed like a mechanical escalator but rather like a leaping dragon. As Wally Broecker and others had been warning, it turns out that the Earth's climate system has often changed in abrupt, unanticipated jumps. The Intergovernmental Panel on Climate Change recognized this as established fact in its 2001 report, declaring flatly, "Long-term observations and experimental insights have demonstrated convincingly that *smooth, or regular behavior is the exception* rather than the rule." The IPCC's projections and scenarios, however, do not reflect this kind of discontinuous change. Moreover, the culture at large has not yet digested this new, revolutionary image of nature, which carries profound implications for the human prospect.

Abrupt climate change, it should be emphasized, is not some theoretical possibility. It has happened before and happened repeatedly. But it took decades for scientists to overcome the reigning assumption of gradual change and fully accept accumulating evidence of past climate shifts so radical and swift as to be inconceivable based on their scientific training. Charles Lyell's influence had extended far beyond Charles Darwin and *The Origin of Species*. His notion of a gradual, unvarying pace of change in nature through all time—known as the uniformitarian principle—had served as the foundation for the study of geology, and this belief in slow, steady procession continued to influence the study of past ice ages and climate change through most of the twentieth century. As early as the 1920s, researchers began to find evidence in tree rings that contradicted this conventional wisdom, but when these early investigators of the historical record presented data testifying to more rapid shifts, their pioneering work proved a hard sell to skeptical peers.

The "discovery" of abrupt climate change, as historian of science Spencer Weart has recounted, involved more than four decades of successive giant steps toward the unthinkable. In the 1950s, the most radical suggestion was that major climatic shifts have taken place in *a few thousand years* rather than tens of thousands of years. By 1960, Wally Broecker and colleagues at Columbia University's Lamont-Doherty Laboratory reported evidence of a massive global shift in climate in *less than a thousand years*. And then as scientists explored ice cores and developed various theories and models about how climate could shift that quickly, they found reason to believe that the climate system did not change incrementally but in jumps. So "rapid" climate change now meant *perhaps a century* or, as Wally Broecker came to believe, *even a decade*. Throughout this revolution, the establishment consensus lagged far behind the leading edge of science. In a 1975 report, *Understanding Climatic Change: A Program for Action*, the National Academy of Sciences saw nothing to indicate that climate change would bring anything but gradual, small changes over centuries. As Weart notes, this history shows that "even the best scientific data are never that definitive. People can only see what they find believable." So the aphorism "seeing is believing" seems to have it backward. More often than not, believing is seeing.

By the early 1990s, scientists were reporting a wide range of evidence of abrupt shifts in past climates—in deposits left by retreating glaciers, lake and ocean sediments, tree rings, coral reefs, pollen laid down in the layers of peat bogs, and in other parts of the planet that hold historical records. Yet resistance to this conclusion persisted. Cores drilled in Greenland as early as 1964 appeared to confirm the reality of climatic jumps, but the controversy had continued because of flaws in the annual layers of ice—thin alternating bands of transparent and cloudy ice from snow added in winter and summer—that hold the record of past temperatures and climate. As the ice sheet had flowed over uneven ground, some of these layers had buckled and folded—a disruption that led some skeptics to question the analysis. Many scientists still found it impossible to believe that the Earth system could leap as wildly and quickly as Broecker and others claimed.

Then in 1993, this debate itself finally came to an abrupt end. Spurred on by this controversy, the United States and Europe had fielded parallel expeditions to drill ice cores from sites on the summit of the Greenland ice sheet roughly nineteen miles apart, in the hope of obtaining an indisputable record of past climate. The U.S. team, the Greenland Ice Sheet Project Two, commonly known as GISP2, hit bedrock in 1993 and completed the task of retrieving a record of the Earth's climate history spanning the past 110,000 years. As scientists began to analyze the GISP2 core and the matching core drilled by the European team of the Greenland Ice Project, they found the two told the same story about past climate—a story of shifts, and jumps, and staggers, and stutters, and flickers, and wobbles, and dramatic climatic reversals that had come on or ended with breathtaking speed. Scientific opinion shifted virtually overnight. It was no longer possible to hold on to the old view of an Earth that waltzed slowly in and out of ice ages like a grand and gracious lady. The most mind-boggling insight from the ice cores is that rapid climate change is *normal*; it is the rule. When the Earth system changes, this is how it behaves.

The ice cores drilled from Greenland and Antarctica also tell us that we live at a truly extraordinary time within this long, volatile climate history, a rare period blessed with a warm and stable climate that has now lasted almost twelve thousand years. During a visit to the National Ice Core Laboratory in Denver, I sat for a long time contemplating a graph with a red line tracking temperatures in Greenland through the most recent ice age and the interglacial period we now live in—a period of roughly 110,000 years. This line surges like a roller coaster through great peaks and valleys of coldness for over a hundred millennia and then soars upward in fits and starts and reversals and renewed ascent to our own time, the long summer since the last ice age, known to scientists as the Holocene. Then the sweeping temperature excursions simply stop, and the red line settles into a dense scribble stuttering within an extremely narrow range of climatic possibility. The difference in this climate record between most of the time in recent Earth history and our time is positively stunning. It looks as if this immensely dynamic climate system had suddenly fallen asleep for the duration of the long summer.

For the past million years, the Earth has swung between ice ages and these warmer interglacials, but moments on Earth with a climate as warm and benign as ours today have been few, fleeting, and very far between. The warm spells recorded in the slender tubes of ice seem brief punctuations between lengthy, fitful, icebound ages lasting 100,000 years; over the past 430,000 years, they amount to 10 percent or less of this long span of time. The three interglacial respites before ours lasted no more than 6,000 years. The only other time in this ice record when the climate stayed so mild and steady for so long occurred 410,000 years ago—long before modern humans appeared on the scene—and lasted 28,000 years. During this time, the Earth's orbit, which changes shape over time and plays a role in climate cycles, was similar to its orbit today, so it provides a reasonable analogue for the natural course of our own interglacial, a course now being disrupted by human alteration of the atmosphere and climate system. Without this interference, this mild time might have lasted another 10,000 to 20,000 years.

This long summer has been critical to recent human history. "It seems unlikely that human societies could have evolved to their impressive level of today in interglacials of 6,000 years or less . . . ," observes James White, who studies the Earth's climate history at the University of Colorado. "We have needed this long period of stable and warm climate to develop modern, complex societies." This unusual climatic interlude has made our current way of life possible. Although it is a distinctly minority view among climate scientists, a few researchers have been trying to make the case that early human agriculture has played a role in extending the long summer by increasing the carbon dioxide and methane in the atmosphere. But whether this saved us from the onset of the next ice age is controversial. Given the similarity between conditions today and the last very long interglacial, observes White, "perhaps we should not be expecting an ice age any time soon anyway." The important point, he notes, is that this extraordinary period is drawing to a close—not because of the advance of another ice age, but rather because of the impact of the modern human enterprise.

The ice cores also tell us that global warming is a misleading term for the danger humans are now courting as we interfere with basic planetary processes. The most profound danger as we push the atmosphere far beyond its normal operating range isn't heat, though increasing heat is no small concern, but rather climate change, which normally arrives in discontinuous leaps. We are now confronting a planetary emergency because this long, tranquil summer that made modern, complex societies possible is coming to an end. We have entered a time of ongoing change, likely surprises, and instability, a time when an angry beast may strike back and "sleeping monsters" may reawake. White, who was one of Broecker's students, protests calling the climate system either "angry" or "a beast" because it casts the Earth system as an adversary when it is simply behaving normally. A worthy point. But these animated metaphors communicate a critical fact to those who meet the planetary era with inherited modern assumptions of passivity, gradual change, good behavior, and admirable balance. The nature we inhabit can be active, powerful, and dangerous. It must be taken seriously.

As scientists have resorted to the language of myth and metaphor to convey the terrifying dangers of the planetary era, I have often thought of the rampaging dragon that ended the life of the great mythical hero Beowulf. I once spent a year learning a vanished language in order to read this ancient poem, which is one of the oldest works in the English literary tradition, and when I had the chance, I visited the British Museum to look at the single remaining manuscript with my own eyes. A similar impulse took me to the Ice Core Laboratory. I wanted to see firsthand the ice cores that contain part of Earth's long, eventful history, the ongoing planetary saga stretching back into deep time, the great story into which every living thing is born.

After the great victories in his youth over the monster Grendel and Grendel's mother, Beowulf eventually becomes the king of his people and enjoys a tranquil reign through "fifty winters" until a thief invades an underground vault where a sleeping dragon guards a vast treasure. The thief makes off with a gem-studded goblet and in doing so, the poet tells us, unleashes something horrible and never intended. Awakened, the dragon "hurtle[s] forth in a fiery blaze" . . . and soon "everywhere the havoc he wrought was in evidence." In a climactic battle, Beowulf kills the dragon with the help of a steadfast warrior but then succumbs to the venom injected with the dragon's bite.

The talk of beasts and monsters expresses the foreboding that we may be unleashing wild, primeval forces in a world that we have recently discovered to be dangerous beyond all imagining. Will our own theft of fossil fuels from the Earth's subterranean vaults awake the sleeping monsters in the Earth system? Will climate change hurtle forth like a leaping dragon? And what kind of havoc will it bring?

Scientists have no way to answer pressing questions about how climate change will unfold over this century, but what they know about the past makes them worried. The Greenland ice cores hold a clear record of the abrupt climatic change that I first heard Wally Broecker describe some two decades ago. A stunning episode some 12,900 years ago known as the Younger Dryas has left "unequivocal evidence" of radical and swift climatic leaps in the not-too-distant past. After roughly 100,000 years of cold and ice, the last ice age finally gave way to a spectacularly sudden warming about 14,700 years ago, and the ice sheets covering North America and Europe began a melting retreat. The transition toward warmer interglacial conditions continued for eighteen centuries. Then, the progress suddenly halted, and the Earth plunged back, within a few generations, to near-glacial conditions in the Younger Dryas, named after a beautiful and rugged white-petaled flower, commonly known as mountain avens, that endures the extremes found in high mountains and the Arctic. Cold, dry, windy conditions persisted for twelve centuries before this interlude ended in an abrupt warming during which the average temperature in Greenland jumped by as much as *18 degrees F.*

The swift warming around 11,700 years ago delivered a climatic jolt felt in North America, Europe, the Middle East, and other regions already inhabited by modern humans. Our ancestors lived through these harrowing climatic reversals. If such radical climate change were to take place in this now densely populated part of the world, scientists warn, it would "devastate modern civilizations."

Based on the events of the Younger Dryas and other abrupt shifts in the Earth's climate history, it is folly to assume that the change ahead will be steady and gradual. But beyond that, the past can offer only limited guidance. The explosive fossil-fueled growth of the modern industrial era is propelling us into an unprecedented greenhouse world, a super interglacial that is far beyond conditions recorded in the ice-core chronicles. Past leaps like the sudden shifts in and out of the Younger Dryas have usually taken place during cold periods, but warm periods have also seen dramatic shifts. The world was fully in the current interglacial 8,200 years ago, and temperatures in northern regions were a bit higher than today, when temperatures in Greenland plummeted again by 10 degrees F over a century.

Scientists understand a good deal about the Younger Dryas, but they have puzzled over exactly what set off the deep freeze. Wally Broecker had long theorized that a vast reservoir of meltwater at the edge of the retreating ice sheet, known as Lake Agassiz, had found an outlet through Lake Superior and launched a pulse of freshwater down the St. Lawrence River into the North Atlantic, which shut down the ocean conveyor circulation. But new geological studies have recently raised serious doubt about whether this was indeed the trigger. Why the Younger Dryas suddenly ended has also been a mystery.

In 2008, the ongoing investigation of Greenland ice cores yielded new, detailed, and, indeed, astonishing evidence about the past that might finally begin to explain how and why climate can change in leaps. The reason, researchers now propose, lies in fundamental reorganization in the circulation of the atmosphere that can occur in *one to three years*—a switch so sudden, noted the leader of the North Greenland Ice Core Project, Dorthe Dahl-Jensen, that it is "as if someone has pushed a button."

This new analysis indicates that the Ice Age ended abruptly 14,700 years ago "within a remarkable three years," initiating a rapid warming of 18 degrees F that happened in two major spurts over fifty years. The return to the deep cold of the Younger Dryas 1,800 years later took place over two centuries. Then, after more than a millennium, the deep freeze ended about 11,700 years ago with another rapid shift, this time over the span of sixty years, and a temperature rebound of 18 degrees F.

Although the ocean conveyor belt plays a role in this, according to the new theory, its action is only one element in a larger sequence of events that triggered abrupt warming. When Greenland was in the grip of cold periods, the ocean conveyor circulation slowed, leaving heat that would otherwise have been transported northward to accumulate in the tropics and the Southern Hemisphere, warming the oceans. This warming continued until it reached a threshold and suddenly precipitated a shift northward of the Intertropical Convergence Zone—the low-pressure belt that encircles the Earth near the equator and acts as a barrier to the mixing of air between the two hemispheres. And this shift in turn set off "a complete reorganization of the mid- to high-latitude circulation almost from one year to the next," causing rapid warming in the North Atlantic region.

In a paper in *Science*, this team warned that state-of-the-art climate models capture "neither the magnitude nor the abruptness" of these climate leaps. University of Colorado climate scientist James White, a member of the seventeen-member research effort, describes the implications bluntly: "Such a rapid climate change would challenge the most modern societies to successfully adapt." While this analysis provides an explanation for what brings on abrupt warming events, White says questions remain about the onset of the Younger Dryas deep freeze and why it lasted more than a thousand years.

As the climate system shifts from one mode to another, it tends to "flicker" for several decades before settling down. Such flickers—wild swings back and forth between one climate state and another—would likely have profound consequences for agriculture. Even without a jolting leap and such flickering, there is a danger that warming could trigger a return to a pattern of much greater climate fluctuations that would also put agriculture and a civilization built upon it in jeopardy. Despite devastating floods that have wiped out communities and catastrophic droughts that have brought down civilizations, the past 11,700 years since the end of the Younger Dryas nevertheless rank as a time of extremely low climate variability. Before this calm period, our ancestors faced a far more erratic and demanding climate marked by fluctuations from decade to decade that were *ten times greater* than current climate extremes. It is this decade-to-decade variability that has the greatest impact on human life and societies, according to climate specialist and author William J. Burroughs. Living with such extreme swings would be "immeasurably more demanding" and would require "an extraordinarily adaptable, flexible, and migratory lifestyle to adjust to changing environmental conditions." If the climate were to shift back to a regular pattern of such wild extremes, the present form of agriculture, which feeds almost 7 billion people and supports complex civilizations, would not be possible. And settled life in cities like Boston would become simply a memory of a golden time.

While much remains uncertain, one fact is, however, becoming increasingly clear. Change in the real world has been happening faster than the model-based projections in the IPCC reports. This is perhaps not surprising, for scientists who study past climate change have been warning for years that the model simulations cannot reproduce the rapid climate shifts that we know occurred in the past. The models do not adequately capture the nonlinear behavior that is characteristic of the Earth system, the propensity for a relatively slight nudge to the system to evoke—as was the case with the ozone hole—a disproportionately huge response. The 100,000-year-long cycles of ice ages and interglacials are a classic example of this sensitivity, as slight changes in the solar radiation—caused in part by periodic shifts in the shape of the Earth's orbit around the sun—trigger the huge climatic shifts between ice ages and warm periods.

This happens because the Earth responds in ways that amplify the cooling or the warming and drive it to extremes. For example, with a bit less solar radiation, more ice forms, and this ice reflects more sunlight back to space, which leads to cooler temperatures and even more ice and more cooling. The reverse effect happens with a bit more warmth—melting and open water capture more warmth and accelerate melting and warming.

The alarming miscalculation regarding the stability of the world's major ice sheets is a good example of the gap between model-based forecasts and real-world change. In its 2001 report, the IPCC had concluded reassuringly that disintegration of the ice sheets was not an immediate concern, for the models of glacier flow indicated that they would be stable over the coming century, perhaps for another millennium. In the time since then, scientists have discovered to their shock that the glaciers in Greenland and Antarctica aren't simply melting; they are cracking and shifting and slipping and sliding faster and faster toward the sea and could be heading for a catastrophic disintegration in the near future.

One glaciologist explained the surprising developments in Antarctica and Greenland, saying, "We didn't know the process; I think we're seeing it now. And it's not gradual." NASA's James Hansen anticipates that this accelerating loss will hit a threshold, and the ice sheets will break up in an "explosively rapid" collapse that will launch "great armadas of icebergs" into the oceans and begin the surge of sea levels. No one knows how close we are to this threshold or whether we might have crossed it already, but the ice sheets are on the move and picking up speed.

If the ice sheets collapse, the world may again face the kind of astonishing surge in sea levels that happened 14,500 years ago. Over the 400 years that followed, sea levels rose an average of 1½ feet a decade—an unimaginably fast rate. By the time the surge ended, sea levels had climbed 65 feet. The last time the Earth reached the temperatures anticipated conservatively by the end of this century—3 to 5 degrees F warmer—sea levels were 82 feet higher than today. Beacon Hill, the highest spot in Boston's downtown, rising 100 feet above the harbor, will become a tiny island. The rest of the historic district with its revolutionary landmarks will go under along with New York, London, Sydney, Shanghai, Tokyo, Calcutta, Bangkok, Bangladesh, most of Florida, much of the Netherlands, the Nile Delta, many Pacific island nations, and coastal cities around the world. Such a huge, rapid jump in sea levels would drive at least a half billion people inland. Hollywood would be hard-pressed to exaggerate the likely chaos.

Most of the dramatic new evidence indicating rapid, perhaps runaway, change does not appear in the 2007 IPCC report, because a conservative review process excludes the controversial, the not fully quantified, and any research not yet incorporated into the climate models. Furthermore, British researchers contend that the government agents involved in the IPCC process significantly watered down the scientists' draft of the "Summary for Policy Makers," consistently removing or softening references to possible acceleration of climate change. One accredited reviewer of the report, David Wasdell, published a line-by-line analysis on the Web comparing the scientists' draft and the final report that supports this contention.

So the worst-case scenario in the IPCC summary for policy makers is a 2-foot rise in sea levels in the coming century. This estimate is based on current models that still assume ice sheets melt slowly rather than break up dramatically. Those who have been watching the astonishing changes on the ice sheets think the coming century may see at least 6 feet of sea-level rise or, in the event that the West Antarctic Ice Sheet collapses, even 18 feet.

At the heart of our dilemma lies ignorance. Scientists cannot predict with any accuracy or confidence how the climate system will respond in a world where carbon dioxide levels are more than twice what they were at the start of the Industrial Revolution, to say nothing of three or four times greater, as may well be the case if the human enterprise continues on the current trajectory. *Depending on how the earth responds—* scientists call this *climate sensitivity—*the future could be anything ranging from seriously disrupted to utterly catastrophic.

Ongoing climate research has not narrowed this range in large part because of great uncertainties about how clouds will figure into the heat balance of the planet as the warming proceeds. Low-lying clouds might block the sun and offset some of the greenhouse effect, while high wispy clouds may prevent the escape of infrared radiation from the atmosphere to space and intensify warming. For more than two decades, cloud behavior has loomed as the weakest link in predicting future climate change, and if anything, the cloud question now appears even more uncertain. By one estimate, the role of clouds in controlling the Earth's temperature could be as much as four times greater than previously thought. When researchers have incorporated this greater range of cloud behavior into standard climate models, the outcomes included hair-raising possibilities. In 2005, the journal *Nature* published a new collaborative study, organized with the help of Oxford climate scientist David Stainforth, that made the case for a more cataclysmic future than the current worst-case IPCC scenario— 9 degrees F warming—anticipates: Stainforth found that there is an outside chance that a doubling of atmospheric carbon dioxide would take us into an unimaginable world where the global average temperature would be 20 degrees F hotter than it is today. Work based on new assumptions about cloud behavior done at Britain's Hadley Centre for Climate Prediction arrived at similar conclusions. Catastrophic warming is not the most likely outcome from doubling levels of carbon dioxide in the atmosphere, but it cannot be ruled out. It all depends on how clouds respond to the warming.

And clouds are just one of many pressing questions. Just recently, scientific teams in the United States and in Europe published studies that also conclude that the warming could be greater and faster than anticipated because climate models have not taken sufficient account of responses in natural systems set in motion by rising temperatures— responses that will amplify the warming. Other studies have found signs that the land and ocean processes that soak up excess carbon dioxide are losing steam. In recent decades, only half of the carbon dioxide created by burning fossil fuels has remained in the atmosphere. The rest has been taken out of the air by plants on land or by the oceans, where some of it dissolves into surface water, converts into bicarbonate, or gets taken up by phytoplankton for photosynthesis or the construction of calcium carbonate shells. Scientists had not been expecting the slowdown of these processes until the second half of the century, but some carbon reservoirs are showings signs of what could be overload decades sooner than expected. One study concluded that because of the unusually hot, dry conditions since the turn of the millennium, plants have absorbed less carbon dioxide. Another research team reported that the North Atlantic has recently been taking up only half as much as it did in the mid-1990s. Even more alarming, a study looking at the Southern Ocean around Antarctica, which is the other major oceanic carbon sink and a key region regulating global climate, found evidence that the waters there were already saturated and unable to take up much more carbon dioxide from the atmosphere. In late 2008, however, further research found that this critical carbon sink had not shut down; the currents around Antarctica had shifted position, but not slackened or changed. For the time being, these waters are still soaking up carbon dioxide and sparing us the full impact of the excess carbon dioxide that human activities inject into the atmosphere.

A greater understanding of the dynamics of complex systems has diminished any hope that one might reliably predict how nature will respond to continuing assaults. If we cannot predict, then we had better beware of the dangers awaiting us in the uncharted territory of the planetary era. Instead of focusing on the smooth curves of escalator scenarios, we need to approach the future mindful of the possible booby traps and wild cards, notably time lags, thresholds, and feedbacks.

Time lags haunt our future. They are a major booby trap of global warming. The slow response time of vast planetary systems puts us in extreme jeopardy, especially given the continuing exponential expansion of the global economy, because of the gap in time between insult and the appearance of injury. This may sound inconsistent with the notion of abrupt change, but it isn't in fact. The Earth system may be slow to respond, but when it does finally react, change may come swiftly. Because of time lags, a generation or longer can pass before planetary systems register the damage inflicted by industrial civilization. The levels of carbon dioxide began soaring dramatically in the middle of the twentieth century after World War II, but the first signs of what appeared to be global warming—successive years of record-setting global average temperatures and the bleaching of corals as ocean temperatures exceeded their tolerance—did not emerge until the late 1980s. The accelerating change we are now experiencing is a consequence of emissions released in the 1960s and 1970s. Time lags mean that it already may be far too late by the time the first sign of trouble appears to prevent major damage or perhaps outright catastrophe. Time lags and momentum also mean that the warming will continue long after we change our ways.

Even worse, these systems can appear deceptively unresponsive up to the moment that an invisible threshold—a point of no return—is reached. Then, with little or no warning, they can shift abruptly and irreversibly. Some use the term *tipping point* for this phenomenon, although this seesaw image can be misleading. One might see a tipping point approaching, but thresholds are apparent only *after the fact*. We have crossed some thresholds already. It is already too late to prevent global warming, to save the world we have known. Indeed, we may have crossed this threshold long before we recognized the danger of climate change. It is beginning to look as if we also may have stumbled across a fateful threshold with regard to the Greenland and West Antarctic ice sheets. In a classic case of threshold behavior, the ice sheets looked solid and stable until they suddenly heaved into motion and began galloping toward the sea. If they are now headed irreversibly toward rapid and spectacular disintegration, the world will soon witness the first catastrophic surprise of climate change.

Feedbacks are nature's wild cards and an essential feature of the nonlinear behavior of our planet. The Earth system has feedbacks that can act as a brake or an accelerator on change. Over the past 4 billion years, the negative feedbacks that serve as a brake have acted to keep Earth's temperature within the range that allows for the presence of liquid water. Somewhat paradoxically, the Earth system has exhibited an overall stability of this sort that has kept it generally habitable, as well as the roller-coaster variability seen in the temperature record over the past 100,000 years from the Greenland ice cores. This variability and the huge swings between ice ages and warm periods are a product of shorter-term positive feedbacks that amplify slight nudges into leaping change. As we push the climate system, scientists anticipate positive feedbacks that will drive the change faster and faster.

The Earth is already amplifying the crisis manufactured by modern civilization, by responding to global warming in ways that cause even more warming. The same positive feedback that has sent the world every hundred thousand years or so from ice ages into warm spells is now at work in the Arctic. The sea ice over the North Pole is disappearing at three times the rate anticipated because the computer models are missing a key positive feedback related to the reflectivity of the Earth's surface, a property called albedo. As the sea ice over the North Pole disappears, the dark open water retains more of the sun's heat, which the bright white ice would have reflected back into space. More warming, more melting, more open water, and on and on in a vicious circle. Another critical feedback is under way in Siberia, where the average temperature has risen by 5.4 degrees F over the past forty years. A vast peat bog the size of France and Germany has started thawing and blowing off the potent greenhouse gas methane five times faster than previously expected. This great bog contains an estimated 70 billion tons of methane, which could quickly escape and jolt the climate system further. Boggy permafrost soils are melting all over the Arctic, raising the prospect that the *1,600 billion* tons of carbon they contain will be making its way to the atmosphere as either carbon dioxide or methane. How this happens and how fast will matter greatly in the decades ahead.

Contrary to the inherited ideas about nature and human possibility that we've carried into the planetary era, our fate in the twenty-first century and beyond will depend not only on what we do or don't do, but equally on how natural systems respond to the growing disruption. This is the paradox of this new historical situation. It is not simply that humans have become a force that now dominates the planet and disrupts the Earth's metabolism, making this time truly the Anthropocene. This proposed name is a bit misleading because some might take it to mean that humans are now in charge. In truth, in bringing the long summer to a premature close and engaging with vast planetary systems, we have opened the door to nature's return as a major, perhaps decisive force in human history.

It is already too late to prevent global warming. But the fact that we have crossed some ominous thresholds before we fully recognized the danger does *not* mean it is too late to do anything at all. Although the challenges of the planetary era are both daunting and without precedent, humans have had long experience with the vagaries of the Earth's climate system. Indeed, its fitful variability has helped make us who we are.

A Stormworthy Lineage

One day almost two decades ago, I suddenly recognized that, contrary to the conventional wisdom in our scientific society, facts and scientific data don't "speak for themselves." I had just finished filing a front-page story about the findings from an urgent scientific mission to determine whether the cause of the Antarctic ozone hole was natural or human. A bit of the adrenaline began to seep away as I sat motionless at last amid the ruins of the day—a sandwich wrapper littered with flakes of sourdough crust, a half-eaten apple, a stained paper coffee cup, the jumble of background papers and press releases, printed copies of earlier stories I had written, and my reporter's notebook opened to key sections underlined in red. I remember savoring for a moment the delicious satisfaction of a mystery finally solved. Scientists had found the answer: Humans had caused the ozone hole; the culprit was CFCs. Then my mood shifted abruptly. These findings transcended ordinary news. The facts felt full of portent and the deep throb of history. This was without question the most important story I had ever written. I felt a flutter of panic. I didn't know what it meant. I had the scientific facts, but I did not know what it all *meant*.

Questions flooded into my mind. Was this just an unfortunate accident, a slipup, an isolated surprise? Or was there something fundamentally wrong about the current approach to the world? Was there perhaps something wrong with *us*?

These questions, once asked, seemed urgent and in demand of answers.

This was the beginning of a long exploration, which I came to call "in search of honest hope." Shortly afterward I began a file, which has since grown fat with diverse explanations for why we now find ourselves in a planetary emergency. Its contents reflect a valiant effort by successive generations over the past 140 years or so to understand the rapidly changing relationship between humans and nature in the turmoil of the Industrial Revolution and in view of the growing human power to command resources, reshape the Earth, and disrupt the fundamental metabolism of planetary life.

We may pile fact upon fact—historical facts about the confluence of events that made modern civilization into a planetary force or scientific facts about symptoms that put us in ever greater jeopardy—but none of them speak to this deeper question about what the crisis means and how to make sense of the human presence on Earth. These facts need a story, and the story we enlist reflects not only our hopes and our fears, but also our vision of human possibility.

Several years ago, I watched a thoughtful conversation at the Harvard Seminar on Environmental Values descend abruptly into fatalism and despair about humans and the human prospect. The spark was a question from a member of the Harvard faculty that had little to do with the two seminar presentations, given by an Old Testament scholar and a distinguished biologist. The gist was this: "Are humans doomed to self-destruct?"

Yes, the biologist replied without hesitation. Humans are already on the way out. Our demise, she predicted with a strange air of satisfaction, is just a matter of time. That seemed to grant permission. A half dozen voices jumped in.

One called *Homo sapiens* a "weed species." Another declared humans a fatal disease, "a cancer on the Earth." A third summed up humanity with the words "selfish and greedy." Then, as I recall, the biologist interjected a quotation from Nietzsche: "The earth is a beautiful place, but it has a pox called man."

This digression was startling, especially given the participants and the setting. The seminar was an invitation-only event attended by academics, writers, policy professionals, and other accomplished people working on environmental questions. Although radical environmentalists are most outspoken, this despair about ourselves is widespread, and anger can simmer beneath the surface even in mainstream environmental discussions. One sometimes encounters an abiding ambivalence about whether to hope for human survival or for speedy extinction.

I know a good deal about the temptation of such dark thoughts. In my work as a journalist, I have been an eyewitness to the devastating, utterly heartbreaking losses taking place at human hands around the world. I've looked a formidable male orangutan in the eye and tracked logs from the vanishing rainforest moving down rivers in Borneo. I've slogged in stinking, ankle-deep manure from factory farms and feedlots and waded in slick crude oil on Alaska's beaches. I've experienced the breathtaking air pollution in Beijing, which hangs in the air like thick fog and swallows the sun. But at the same time, I have encountered such a wide range of human behavior and cultures along the way that I grew wary of pronouncements about the "species." Despite all that I was seeing, I couldn't give in to this essential fatalism.

Our past bears on the human prospect as much as do the dangers and uncertainties of our future. We have inherited a variety of stories, tales of origin, that seek to justify the present as much as provide an account of the past. Origin stories tell us about our place in the world through an account of how things came to be the way they are. They explain who we are in relation to Nature. The traditional Western origin story in the biblical Book of Genesis tells us that humans came in the beginning from an idyllic Eden and that we enjoy as our birthright a God-given dominion over the rest of the natural world.

As the modern era took shape, however, this original story of loss and a fall from grace evolved into a narrative of progress and the eventual recovery of Eden through human effort. Then, in the nineteenth century, after Charles Darwin put forward his theory describing how all life had descended from a common ancestor, a new, scientific origin story began to displace the biblical account, serving as both a scientific theory and a scientifically based modern myth. In the hands of such popularizers as the English philosopher and political theorist Herbert Spencer, the notion of evolution, though in a version quite at odds with Darwin's theory, served to lend "scientific" support to the modern pursuit of progress and the quest for absolute dominion. In this spirit of progressive evolution, a classic illustration during my school days tracing the ascent of humans showed a series of two-legged apes—each less hairy, more upright, and taller than the preceding—that culminates with a modern man striding confidently forward. From the time our remote ancestors stood up on two feet, it suggested, they embarked on an uninterrupted journey of continuous improvement, onward and upward, expanding human power over nature. This march of progress, inevitable and unstoppable, somehow embodied the laws of nature and led inexorably, as one of Spencer's disciples put it, "from gas to genius." Humans, therefore, don't just have a history; our species has a destiny.

The Romantic movement, which arose in protest against the materialism and rationalism of the Scientific Revolution and the Enlightenment, provided the framework for a competing story—an account of an ongoing decline from an original nature, of an earthly Eden being destroyed by the disruptive human presence. In the most pessimistic version of this narrative, humans are somehow fundamentally at odds with nature, an *inevitable* agent of discord in a natural world of balance and harmony, perhaps an egregious evolutionary mistake. This alternative current within modern culture inspired artists and poets and later gave rise to nature writing and the wilderness movement; its antiprogress narrative resonated with many amid the chronic upheaval loosed by the Industrial Revolution. Writing in this tradition in an early book on climate change, *The End of Nature*, Bill McKibben mourned the loss of a separate and untouched nature that partook of the eternal and was characterized by its "utter dependability." So McKibben writes: "We have built a greenhouse, *a human creation*, where once there bloomed a sweet and wild garden."

The emerging picture of Earth's history could hardly be more contrary to either version of our imagined past. It is a chronicle of tumult that contradicts the inevitability of the progress narrative and testifies that our ancestors emerged from the trials of climate hell rather than from any primordial Eden. Long, long, long before humans appeared on the evolutionary scene, the Earth of the Eocene epoch some 55 million years ago did bear some resemblance to the Eden humans have imagined: The entire planet had enjoyed a relatively stable, warm, and moist climate, allowing tropical forests and plants to extend across much of its surface. But this stable, relatively homogeneous landscape vanished with a broad climatic deterioration that would eventually set the stage for human evolution.

Beginning roughly 50 million years ago, tectonic plates—great slabs of the Earth's crust that float on hotter, more liquid layers below—fractured, shifted, collided, lifted, and over tens of millions of years dramatically remade the face of the Earth. In this process, the Indian peninsula drove its way into the Asian landmass, and the rocky crust at the boundary rose up in the slow-motion collision to form the Himalayas and the Tibetan Plateau. In North America, great movements within the Earth lifted vast parts of the continent upward, creating the Rocky Mountains, the Sierra Nevada, and the Colorado Plateau. This great tectonic remodeling shifted wind and rain patterns and likely set off a precipitous decline in climatic conditions, making the Earth progressively colder, drier, more diverse in its habitats, and far less stable. The climate system became increasingly variable and periodically subject to dramatic change. Around 2.5 million years ago, as the North Pole began to develop ice sheets, the Earth's climate commenced its great excursions between ice ages and warm inter-glacial periods.

Over the past 800,000 years, a critical period in the evolution of our species, those who preceded us managed to survive the global effects of eight ice ages, experienced shifts from tundra conditions to a mild climate in a lifetime, and hung on through the wild reversals of transition periods when the climate systems flickered back and forth between cold and warm conditions every five to twenty years. As scientists have gained insight into the Earth's volatile climate history, it has become clear that our ancestors made their way on an erratic and changeable Earth.

How our own evolution fits into this startling new picture of Earth's history is unquestionably relevant in light of the growing emergency. Rick Potts, a noted researcher in human origins at the Smithsonian Institution in Washington, D.C., emphasizes one striking fact about our species' past: Humans evolved in one of the most unstable and environmentally dynamic spans in the Earth's history. This turmoil is as critical a part of the human story as the tranquility of the long summer, which over the past 11,700 years has allowed for the rise of complex civilizations. Bringing together what scientists have recently learned about the Earth's eventful past and the fossil evidence of human evolution, Potts has developed an illuminating account of our "ecological genesis" within this volatile nature. The ancestors of modern humans first appeared in a tempestuous era 5 million years ago, and our evolution since has been driven, not, as often proposed, by a particular environment, such as the African savanna, or by one particular climate, such as the deep freeze of the ice age, but rather,

Potts theorizes, by growing environmental instability. What has emerged from this instability is a versatile human species for all seasons and climes.

This view of human possibility contrasts sharply with one pessimistic explanation that has been offered for our current dilemma—the mismatch between the Stone Age mind that humans are purported to possess and our twenty-first-century crisis. If this is, in fact, the case, then humans must lack the inherent capacity to meet the challenges of the planetary era. We are handicapped, perhaps doomed, by the limits of our evolutionary legacy.

At the same time, the story now emerging about our origins also bears little resemblance to the optimistic story of the smooth linear ascent of humans in that illustration from my school days. Rather we have traveled a crooked path to the present, a path hewn by the tinkering of nature and repeated evolutionary experiments. Humans are just one of perhaps as many as twenty upright-walking hominin species that evolved in the face of climatic oscillation and shifting landscapes. Our 5-million-year family history is in large part a story of extinction. Today humans are the sole survivors, the only member of this diverse family—which scientists long called *hominids* but have recently renamed *hominins* based on new genetic evidence of relatedness—to emerge from a brutal gauntlet of intensifying climatic extremes. "As sweet as our ownership and as proud as our domination of Earth may seem," Potts reflects, "our existence as a species is a bitter miracle."

There is a great irony at the heart of this new account of our origins in a nature about as fractious as a bucking bronco: The rising instability, particularly over the past 700,000 years, forged the very human talents that have allowed us to become a planetary force and an agent of crisis and instability. At the same time, however, this evolutionary legacy also gives me good reason to believe that humans can—with wisdom and luck—make it through the dangerous passage ahead. Our kind has survived repeated environmental trials, which have made us, in Potts's words, "a foul-weather species." This evolutionary track record testifies to our ability to confront the unprecedented and do the unexpected. We are neither destined nor doomed by anything in our nature.

As the news regarding our planetary emergency has gone from bad to worse with breathtaking speed, however, it has become ever more difficult to remain both honest and hopeful. Sustaining hope sometimes seems like walking a tightrope stretched across the slough of despond. The bad news hits like gusts of wind and increasingly at gale force. It is easy to lose balance and tumble into despair. In black moments, wrestling with the possibility that it might be too late to avoid catastrophe and unimaginable suffering and death, I have found myself reaching for Potts's book *Humanity's Descent* for sober consolation. At such times, I need reassurance that "the strange buoyancy of the hominids is in us, a hopeful heritage of response to environmental dilemmas." This is a story we need to hear at this turning point in the human venture, for it testifies eloquently to the meaning of human resilience.

The capacity to respond to unprecedented challenges is within us, if this is the case; it is part of our evolutionary legacy. But the question remains, Why aren't we responding? The answer, in my view, lies largely in our problematic civilization. The radical cultural experiment of the modern era not only drives this planetary emergency, it impedes the necessary understanding and appropriate response, so we fail to take the necessary action soon enough and fast enough. If it proves impossible now to avoid a twenty-first-century version of climate hell, the coming decades will require all the buoyancy and resilience we can muster.

The thing that first set Rick Potts wondering about environmental instability and evolution was a layer of pink silts at a hot, desolate excavation site in southern Kenya called Lainyamok, "the place of thieves" in the language of the area's Maasai herders. The silts bear witness to a spectacular event some 400,000 years ago: A large lake— larger and older than twenty-mile-long Lake Magadi, which is today the southernmost lake in Kenya's Rift Valley—suddenly dried up completely as the area was gripped by intense drought. With a plan to explore how such dramatic shifts shaped our human ancestors, Potts later found a perfect place twenty-five miles away called Olorgesailie, the site of another ancient and now vanished lake, where the eroded faces of hills and gullies revealed alternating sediments of gray, brown, and white. Like the ice cores of Greenland, these layers deposited year by year held a "crisp record" of the past million years of shifting climate, "an archive of environments" that allowed him to trace the ecology of the region through time. The tiny silica skeletons of lake diatoms, single-celled aquatic algae, in the successive strata record restless impermanence as the lake waxed and waned, held steady for thousands of years, disappeared altogether, and then expanded again in fitful reversals that repeatedly reshuffled resources and altered the game of life until finally great tectonic shifts tilted the land upward and emptied the lake forever. Potts walked up and down the hills, studying the colored layers and contemplating the unmistakable oscillation in the environment and, even more intriguing, signs of increasing variability, suggesting that the climatic shifts became more extreme through time. This is the dynamic, shifting stage on which human evolution played out.

As Potts studied the punctuated developments that eventually gave rise to modern humans within the context of this climate history, he found that each of the three most important steps in our evolution emerged during times of especially intense climatic oscillations. The first of these breakthroughs, bipedal walking, which increased mobility and the ability to exploit scattered and diverse food resources, first appeared in an unstable period from 4 million to 1.9 million years ago. Stone tool making emerged between 2.5 million and 1.5 million years ago, and the large brain found in modern humans evolved during the past million years, which have been marked by the greatest climatic swings in the 5-million-year history of our lineage. There is another pattern as well: Over the course of human evolution, as the silt layers at Olorgesailie reflected, the range of these swings between climate extremes grew *three times greater*, posing ever more harrowing survival challenges.

Our branch of the diverse family of two-legged hominins endured, it appears, because our ancestors set off on a path of flexibility and turned to cultural strategies for dealing with these vicissitudes of nature. When, for example, changing climate forced a shift to tougher kinds of food—hard roots and coarse grasses—some early hominins, such as *Paranthropus*, evolved powerful jaws and formidable high-crowned teeth. Others, however, including those who would become our ancestors, tried a new way of meeting a changing world. Instead of committing to bodily changes like *Paranthropus*, whom Potts describes as a "chewing factory," they pursued what might be described as the Cuisinart strategy, using shaped stones to process food. These stones, which served as extremely versatile external "teeth," held a distinct advantage, for they could be adjusted to shifting food supplies far more swiftly than real teeth. These divergent paths—the internal chewing factory and the external Cuisinart strategy— illustrate the two opposing themes that have played out in the drama of human evolution: "the fittest," who *is* adapted to a specific environment, versus "the most flexible," who *can* adapt to diverse conditions.

The large mammals of East Africa, who shared this landscape with these human family members, responded to environmental shifts in similar ways. A million years ago, a collection of huge, big-toothed, grass-eating specialists dominated the East African grasslands—pigs, zebras, elephants, hippopotamuses, and antelopes that were far more massive than those that live on the savanna today. These "lawnmower species," as Potts calls them, enjoyed great success on this dry savanna until the conditions to which they were splendidly adapted gave way to another period of extreme fluctuation, culminating around 600,000 years ago. By 400,000 years ago, all had died out. The survivors were closely related species of elephants, hippos, zebras, antelopes, and pigs, but more flexible types with smaller bodies and teeth and the ability to shift their grazing to different foods as conditions and settings varied.

This same pattern of survival and extinction is evident in the human family tree. As the climate swung more fitfully between extremes, the conditions for survival changed repeatedly, and those committed to a particular habitat and way of life, the specialists, died out. These included most of our now-extinct kin, who lived only in certain areas of savanna in East Africa, as well as the Neanderthal populations, a form of humans with short, stocky bodies seemingly adapted for ice age conditions who flourished in Europe and western Asia for much of the past 200,000 years. Conventional wisdom has held that Neanderthals eventually lost the evolutionary race because they couldn't compete with the more intelligent modern humans who arrived in Europe some 40,000 years ago. But a leading expert on ancient brains, Ralph Holloway of Columbia University, finds no reason to think this close cousin was mentally inferior to our kind. Using Neanderthal skulls to make casts of the brains they once contained, Holloway found that Neanderthals had brains 20 percent larger than that of the average modern human. Moreover, the shape of the frontal lobe—the part of the brain responsible for complex thought—is as advanced as ours, leading Holloway to conclude they should have had similar thinking ability. Archaeological evidence from Neanderthal sites is also contradicting the caricature that this close cousin was a behaviorally static Stone Age dolt. This line of investigation is providing insights into evolving Neanderthal culture and technology and showing that modern humans have not been the only members of the family capable of innovation.

Paleoanthropologist Clive Finlayson, who has studied the Neanderthal sites in Spain where the species endured the longest before finally dying out sometime after 28,000 years ago, has also concluded that competition had little to do with their extinction. He thinks the explanation lies in the ecological and climatic conditions at the time of their demise and the Neanderthal commitment to a particular way of life. They were specialists adapted to exploiting a particular kind of landscape. When rapid, repeated, and extreme climatic shifts remade their world, they found themselves at an evolutionary dead end.

Neanderthals favored the transition zones between ecosystems that offer rich and diverse resources within a relatively small area. They typically lived at the edge of broken woodlands and hunted by ambush in areas that were a mix of grass, shrubs, and scattered trees. Evidence suggests they used the vegetation for cover as they stalked such prey as red deer, which they killed at close range with bayonet-like thrusts from stone-headed spears. They weathered short-term instability by shifting between various foods available at relatively close hand. Through this way of life, they had managed to persist for tens of thousands of years in the midlatitude region stretching from Portugal to the mountains of central Asia.

Then around 45,000 years ago, the climate in Europe began oscillating rapidly, and as these broken woodlands began to shift to open plains, Neanderthals faced a survival crisis. Their physical makeup could not change fast enough to make them effective in a treeless landscape with far-ranging herds of grazing animals. The heavily muscled Neanderthal body, with its extremely short legs and broad pelvis, was built for the thrusting power required in close-range hunting rather than for the speed and mobility this new landscape demanded.

Body structure wasn't the only problem. Judging by the shape of their inner ears, which is a critical part of the body's balance system, researchers believe they were less agile than modern humans or even than their more primitive ancestors. As open habitats spread and encroached on Neanderthal territories, Finlayson notes that those on the edge of this shifting landscape attempted to develop tools suited to the new conditions, but their handicap in an open landscape requiring mobility is evident in severe wear and tear on their joints and limbs. In the end, Neanderthals, who were well adapted to the terrain they had inhabited and exploited for tens of thousands of years, could not cope with the prolonged instability and the speed of the change that remade the world around them. Finlayson believes they eventually succumbed to the cumulative effect of repeated cold periods that shifted and fragmented the habitat to which they were suited. The last of these brought the coldest, driest, harshest climate in 250,000 years.

One evolutionary pattern predominates in the face of this increasing instability and wider climatic swings, Potts observes: "the survival of the generalist." Again and again, the survivors who emerged from these episodes of climate hell proved to be the mobile and the flexible. Those faced with shifting environments have these two basic choices: to move and, if possible, follow critical resources or to find ways to endure a wider range of environmental conditions. Over time, the human lineage responded through bodily changes that made our forebears able to range more widely: They became significantly taller, larger, and fully bipedal with long, muscular legs suited for running or walking long distances. The adaptable generalists who traveled this evolutionary path also became ever more disengaged from any particular place or set of conditions, so they could endure through environmental oscillations, spread and colonize unfamiliar habitats, and respond in novel ways to new challenges and unfamiliar surroundings.

Sometime between 1.8 and 1.5 million years ago, one notable hominin in our family, *Homo erectus*, escaped the geographical and ecological constraints that limited most of the extinct members of the human family tree as well as our closest living relative, the slender chimpanzee called bonobo. A six-foot-tall, long-legged species with bodily proportions similar to a modern human, *Homo erectus* began leaving Africa and spreading across much of Asia—travels that in

Potts's view "foreshadowed the current state of the human species, the ultimate infiltrator and colonizer of all continents and terrestrial biomes on the planet."

The startling discovery of a three-foot-tall fossil with primitive prehuman features dating to only 18,000 years ago on the island of Flores, Indonesia, in 2003 has raised a flurry of new questions about human family history and how much we have yet to discover about it, including the possibility that some early hominin species may have left Africa long before *Homo erectus*. Since then, a scientific debate has been raging about whether these diminutive individuals, who quickly acquired the nickname hobbits, are diseased modern humans suffering from some type of developmental disorder or a new species, *Homo floresiensis*.

In Potts's view, the significance of our own species' later move out of Africa some 200,000 years ago is commonly misunderstood. While the global spread of humans may represent an untethering from the physical place of our origins, it is *not a release from the environment*. Rather it is a response to its long-term inconstancy.

The evolutionary advances that finally gave rise to modern humans took place in the face of intensifying climate swings over the past 700,000 years. During this time of extreme trial, the ancestors of modern humans evolved larger brains along with new survival strategies and an ever greater capacity to adjust behavior and social organization in differing circumstances. They became increasingly skillful at manipulating their environment: They mastered fire, and through intentional burning began to transform the land around them to better suit their purposes. Wielding fire, our ancestors could drive game, scare off predators, clear away undergrowth to make it easier to harvest fallen nuts, and create favorable conditions for plants used for food, medicine, and fiber or for the tender new growth that attracts deer and other game. The ability to control fire also led to the invention of cooking, a cultural step that likely played a critical role in the evolution of larger brains.

Amid the dizzying diversity of life, animals with big brains are rare. If a large, complex brain has obvious advantages, why aren't they more common? Primarily, according to a leading authority on brain evolution, John Allman of the California Institute of Technology, because a big brain is very expensive to develop and maintain. Parents must devote a great deal of time and energy to rearing large-brained offspring, who are dependent on adults and develop slowly. Moreover, once mature, a large brain must compete with other organs for energy, a considerable evolutionary problem. The practice of cooking food allowed humans to transcend this energy constraint. Heating food has several benefits: It kills dangerous bacteria and parasites, breaks down poisons in plants, and predigests protein. Thus, cooking not only increases the safety and the variety of food, it makes digestion far easier, and with less energy required for digestion, more became available for maintaining the brain. In humans, the brain represents only 2 percent of body weight, but it commands roughly 20 percent of the body's energy. In the period between 500,000 and 100,000 years ago, when hearths and fire pits appear in the archaeological record, human brain size increased by 24 percent.

In *Evolving Brains*, a sweeping survey of brains from bacteria to humans, Allman asks an even more fundamental question: Why did brains evolve? "In the broadest sense," he concludes, "brains are buffers against environmental variability." One can see this principle at work in two closely related species of New World monkey, the fruit-eating spider monkey and the leaf-eating howler monkey. Although the monkeys are roughly the same size, the spider monkey's brain is almost twice as large, because feeding on fruit, which is widely dispersed and available at different times, is far more challenging than dining on leaves, which are always within easy reach. The first true brains in the history of life—found in the early chordate *Haikouella lanceolata*—originated amid extreme climate instability in the early Cambrian period more than half a billion years ago. Similarly, the two major expansions in the human brain, first 2 million years ago and then beginning 500,000 years ago, are linked to the challenges posed by periods of extreme habitat variability.

These large brains and the path of increasing flexibility gave rise in time to the unique behavior of modern humans—food sharing, cooperation, extended families, and social networks as insurance in the face of uncertainty; the ability to innovate and use tools to make new tools; language and the ability to transmit information about things not immediately visible; and the cultural capacity that allows for many different approaches to living and makes us a species characterized, says human-origins researcher Ian Tattersall, by its "bewildering variety." Like other important leaps, the cultural creativity of modern humans—which was unquestionably evident by 40,000 years ago, in the use of bright paints for body decorations and rituals, and somewhat later, between 32,000 and 15,000 years ago, in the flowering of art in carvings, sculptures, painting, and intricate jewelry—emerged in the midst of extraordinary climatic alterations. Human culture does not, as has been long assumed, stand in opposition to nature. Rather, Potts concludes, it flows directly from nature and the turmoil of our evolutionary past.

There is an inclination, particularly in the United States, to underrate the importance of culture in explaining human behavior and to blame current self-destructive trends on our genetic endowment. The fatalism I encountered at that Harvard seminar is not uncommon. The *real problem*, I've been told many times during the question-and-answer period following presentations, is that "humans are basically selfish and greedy." The assumption that all humans share a single, fixed nature is widespread and, in the view of the eminent evolutionary biologist Paul Ehrlich, highly problematic. Nothing is more in error, he says, than the notion that "people possess a common set of rigid, genetically specified behavioral predilections that are unlikely to be altered by circumstances." The old saying "you can't change human nature," he says, is flat-out wrong. The truth is, humans are impressively change-able, which explains why our kind can exhibit a stunning diversity of "natures" shaped as much by evolving cultures as by changes in the ge-netic information passed on from our ancestors. If we are going to solve the problems humanity now confronts, he says, we have to begin with a proper understanding of ourselves and an appreciation of "the overwhelming power of cultural evolution—the super-rapid kind of evolution in which our species excels."

The struggle for a proper understanding of ourselves has centered on this puzzle of biology and culture, the perennial nature/nurture de-bate, which in its most caricatured form has been a standoff between those who contend that humans are largely hardwired by instructions in their genes and others of the opposing blank-slate school, who argue that humans are infinitely malleable. Most reasonable people, however, would probably accept the moderate and sensible opinion that both biology and culture help shape who we are, but the assump-tion has generally been that culture and nature are *opposing* influences. Some, looking through the evolutionary lens, view culture as an arti-ficial force that suppresses the natural, one that censors and frustrates our animal instincts and thwarts aspects of our own nature. In this view, culture may dictate monogamy, while people's animal selves in-cline them toward adultery; thus culture goes against the grain of human instinct. Others have seen culture as a redeeming aspect of human existence in that they believe institutions like religion help us transcend an undesirable, irrational, animal part of our nature. Either way, nature and culture are considered to be essentially at odds.

Now, however, groundbreaking work in brain science and psy-chology is showing that biology and culture work in concert rather than in opposition, revealing how this creative collaboration results in the unique, flexible, diverse human ways of being. Scientists at work in the emerging field of cultural biology have been using brain imaging, computer modeling, and advances in genetics to illuminate the dynamic interplay between brain and culture, which have coevolved and given rise to the special talents that set modern humans apart from their predecessors. Steven R. Quartz of the California Institute of Technology and Terrence J. Sejnowski at the University of California at San Diego, pioneers in this investigation that traverses disciplinary boundaries, describe the human brain as anything but hardwired. At birth, the brain is an unfinished organ that is actively constructed over years through its interactions with the human and natural environment. Making a human requires a partnership of genes and culture—the uniquely human social and cognitive phenomenon that allows for the sharing and transmission of information and ideas across numerous minds and through time and space. With language and the written word and now other technologies, it is possible for humans to do such wondrous things as share directly in the thoughts of those long dead and send messages to those yet to be born. Culture is in some sense participation in a larger, communal mind, a shared system of meaning that forms the basis for cooperative action and the development of enduring human institutions and complex civilizations.

Proper human brain development *depends* on interactions with culture. "It's not a question of nature or nurture," according to Quartz and Sejnowski, "but an intersection of the two that is so thorough that it dissolves that simplistic dichotomy." The brain and the world in which it develops is "an integrated system, a fabulously tangled web of connections." Accepted dogma has impeded brain science as much as it has climate science. Just as climate scientists adhered to the belief in gradual change, those in neuroscience long accepted without question the notion that the brain did not add new cells after its initial development. Those of us who were young in the wild days of sex, drugs, and rock and roll in the 1960s became acquainted with this supposed finitude of brain cells through advertising campaigns intended to dissuade us from at least one of these vices. Smoke a joint and you might kill off a few more of your limited lifetime supply of brain cells—or so the scare story went. But this notion of the static, finite brain also turned out to be utterly untrue. The brain is dynamic and constantly changing, and as it engages with the world, it can grow new cells throughout life—a new view that has emerged from a dramatic revolution over the past decade or so in the understanding of how the brain develops.

In the old view, it was thought that the brain as a whole developed extremely rapidly in the first two years of life, producing a great excess of brain connections, the maximum a person would ever have, and then somehow this surfeit got progressively whittled away during further cognitive development. Now, however, it has become clear that brain development takes place in stages over a far longer time and that different parts of the brain develop at different rates. Quartz and Sejnowski believe this hierarchical staging of development allows the human brain to engage in what they describe as "constructive learning," interactions with the surrounding environment that help wire its circuits. Learning isn't, therefore, just a matter of storing information in the brain, but rather a process that actively shapes the brain's *physical structure*. One often hears the brain compared to a computer, a metaphor popular among cognitive scientists, but Quartz and Sejnowski note that a computer does not generate new circuits as you record information by typing on its keyboard. But in a human, learning *changes* the brain's hardware.

Even more intriguing, Quartz and Sejnowski theorize that this strategy of constructive learning helps explain how humans managed to evolve a brain of such complexity and why "flexibility lies at the core of who we are." If one compares the genes in a human to those in a bonobo, the difference between the two species amounts to a handful of genes—1.6 percent, to be exact. There is, in fact, a greater genetic difference between a bonobo and a gorilla. The puzzle has been to explain how, with so little genetic difference, the human brain manages to have a vastly greater number of neuronal connections and vastly greater structural complexity than that of our closest primate relatives. The prefrontal cortex—the part of the brain behind the forehead where many uniquely human capacities, including a sense of self, appear to reside—is six times larger in humans than in another primate cousin, the chimpanzee. But there are similarities as well as differences, including reports that some chimpanzee groups pass on learned behavior, such as making tools used to fish for termites. Although animal behaviorists rightly describe this as simple culture, chimpanzee brain development does not depend on culture in the way human brain development does. Chimpanzees do not become who they are through this kind of constructive learning.

Beyond the high energy requirement noted by Allman in his analysis of why big brains are rare, the forces pushing human evolution toward larger and more complex brains encountered two other obstacles: a limit to how much genes can specify and a limit on the physical size of brains. So if the brain was going to become more complex, it would be necessary to find a way around these limits. Our species managed this feat, and, most fascinating of all, we did so through a strategy seen before in human evolution—by *externalizing* the process of brain development, in part, through culture. Just as the path of flexibility saw stones used Cuisinart-like as external teeth and cooking employed to externalize part of digestion, the development of the human brain has come to rely more and more on the outside world—especially on the memory and information carried in the surrounding human culture—rather than solely on some precise internal program specified in our genes. So while birds of some species are born with an instinctive "map" specified in the genes, a built-in behavioral program that guides them on a long migration, the famed Polynesian and Micronesian navigators who ranged across the open seas of the Pacific and colonized remote islands accomplished their long voyages by relying instead on traditional knowledge of the currents and winds and a cultural legacy of detailed star maps. Once human evolu-tion set off on this path, the process proved self-reinforcing. As cul-ture came to play a greater role in human brain development, the brain grew in size and complexity, which enabled more extensive cultural practices, including the development of language and symbolic ac-tivities, which in turn helped make the brain become even more complex.

This kind of development through dialogue with the world pro-duced brains that were both more complex and more sensitive to the environment. This *"progressive externalization"* of the brain's develop-mental program, Quartz and Sejnowski conclude, was "a way to make the brain more adaptable to an uncertain world." Like Potts and All-man, these pioneers in cultural biology regard this evolutionary leap leading to modern humans as a response to instability, particularly the wild climatic swings over the past 700,000 years. These vicissitudes made us who we are—"the most complex collaborative project in his-tory" involving two equal partners: "our biology and the human cul-ture we are immersed in."

This collaboration of culture and biology explains how the "bewilder-ing variety" in human behavior comes about. Moreover, cutting-edge research in the field of psychology is neatly dovetailing with Quartz and Sejnowski's work in neuroscience. Cross-cultural investigations have been exploring the reality of "human natures" and detailing pro-found differences in the way people from different cultures think and view the world.

Americans and their leaders seem particularly disposed to the as-sumption that cultural differences are superficial and that once one pushes beyond this cultural veneer, everyone in the world is essentially alike. This universalism reflects the legacy of ideas from the European Enlightenment and gives rise to the naive conviction that the world would be better off if other nations simply adopted Western institutions and values. As the United States under President George W. Bush pursued a foreign policy based on this stated objective, the British historian John Gray described it as "the world's last great Enlightenment regime."

Science, which is part of this same rationalist Enlightenment tradition, has also been under the sway of universalist assumptions, including the belief that all humans perceive and think essentially the same way. Culture might tint the lens and shift the focus a bit or teach people different things, but cognitive scientists have generally taken it for granted that the basic cognitive process does not vary. This widespread assumption has gone hand in hand with the belief that the brain was for the most part hardwired according to a genetic blueprint.

Psychologist Richard E. Nisbett of the University of Michigan was himself "a lifelong universalist concerning the nature of human thought" until a fateful encounter with a student from China set him off on a new course of groundbreaking research that eventually upended everything he had believed. What Nisbett discovered gave rise to the new field of cultural psychology—a name he chose to reflect the incorporation of culture into the scientific study of cognition. Some of this work comparing thought processes in American and Asian students found striking differences in the way the two groups see and think, differences rooted in culture. Asians and Westerners do not merely learn different things and hold opposing opinions; they use *different tools* for understanding the world, a discovery that supports the view that culture does participate significantly in the wiring of the brain.

The challenge to the universalist assumption upon which Nisbett had based his career as a psychologist—that all human groups think and reason the same way—came from his student Kaiping Peng. "The difference between you and me," Peng told Nisbett, "is that I think the world is a circle and you think it's a line. The Chinese . . . pay attention to a wide range of events; they search for relationships between things; and they think you can't understand the part without understanding the whole," he argued. Westerners, on the other hand, "think you can understand the part without understanding the whole; they focus on salient objects or people instead of the larger picture; and they think they can control events because they know the rules that govern the behavior of objects."

Nisbett was stunned, but at the same time open-minded and intrigued. Soon he embarked on cross-cultural research with colleagues at universities and institutes in Beijing, Kyoto, and Seoul to examine exactly how people born into different cultures see the world, remember, and think. This extensive laboratory investigation showed repeatedly that thought processes and perception differ profoundly between East Asians and Westerners, largely in the ways Peng had described. What explains two utterly different cognitive systems, so opposite in their thinking, views about the nature of the world, and social relations? The answer, Nisbett concluded, was culture, which has perpetuated these distinct approaches for at least 2,500 years. These thought systems—already evident in the teachings of the Chinese sage Confucius, who lived from 551 to 479 B.C., and in the philosophy of the Greek philosopher Aristotle, who lived from 384 to 322 B.C.— have persisted for millennia through cultural dynamics that create self-reinforcing, homeostatic systems. Thus, he says, "social practices promote worldviews; the worldviews dictate the appropriate thought processes; and the thought processes justify the worldviews and support the social practices."

The Western emphasis on individual action and the Chinese focus on relationship are even evident in the primers that have been used to teach children to read. The Dick and Jane book used in my first-grade class begins with an individual in action: "See Dick run. See Dick play. See Dick run and play." The Chinese counterpart from the same period shows a big boy with a little boy on his shoulders: "Big brother takes care of little brother. Big brother loves little brother. Little brother loves big brother." Nisbett also cites research showing similar differences in the way Japanese and American mothers teach their children: The patter between Japanese mothers and their babies tends to emphasize relationships, while Americans tend to point out and talk about objects, such as dogs and trucks. Nisbett reports that American infants learn nouns most rapidly, while Asian infants learn verbs—a pattern rooted in these child-rearing practices that help perpetuate culture.

As adults, Americans have the sense of personal agency that was so characteristic of the ancient Greeks and a conviction that the world can be controlled. Asians, on the other hand, see the world as a far more complex place, which is not easily controlled. They are more likely to adjust to circumstances than try to change them. It must be noted, however, that the Maoist Revolution and Marxism, a quintessentially modern faith, injected a strong dose of Western ideas and aims into the thinking of the Chinese leadership. While the traditional Chinese aspiration had been to find harmony with nature, leaders in the Communist era have pursued the aggressive conquest of nature, a posture that China's rapid industrialization and entry into the global economy may further reinforce. The impact on the culture at large, however, is not yet clear.

This provocative work by Nisbett and his collaborators leaves little doubt that the differences between Asians and Westerners are large and go much deeper than particular beliefs. The testing has shown again and again that people from the two cultures actually *view* the world differently. When people from the two cultures look at the same picture, they see different things. They also reason and explain the world in very different ways. Asians have "wide-angle vision" and take in the context of a given scene; Americans, on the other hand, see with "tunnel vision" that tends to focus on prominent objects and be inattentive to context. So, explains Nisbett: "People with a wide-angle view might be inclined to see events as being caused by complex, interrelated contextual factors whereas people having a relatively narrow focus might be prone to explain events primarily in terms of properties," such as the inherent nature of the people involved. Testing in cultural psychology has shown this to be the case: Asians are far more likely to judge that the surrounding context influences a person's behavior, while Americans tend to attribute the behavior solely to one's personality or character or genetic makeup.

The world seen through Western eyes is a far simpler place. When confronted with contradictions, for example, Westerners are inclined to judge that one belief is correct and the other wrong. According to Nisbett, "Americans' contradiction phobia may sometimes cause them to become more extreme in their judgments" in the face of contrary evidence. Rather than moderating their opinion, they become defensive and more adamant about their original view. Asians, who think dialectically rather than analytically, find it easier to see some truth in both sides and seek "the Middle Way," as encouraged by Buddhism. They do not feel the need to choose one side or the other and are not prone, in the way Westerners are, to dichotomous black-and-white thinking that excludes shades of gray.

As Nisbett searched for an explanation for these contrasting approaches to the world, he found reason to believe that they grew initially out of differing physical circumstances in ancient Greece and China, that cultural patterns emerged in response to ecological and economic conditions. The fertile plains, navigable rivers, and low mountains of China favored agriculture, and the irrigation systems that developed in such places as the Yellow River Valley required centralized control of the society. Constructing and maintaining such extensive irrigation systems are impossible without a significant degree of harmony and cooperation. Western culture emerged out of a dramatically different ecology. In the Greek landscape, where mountains descend to the sea, people made their livings through hunting, fishing, herding, and trade, occupations that require far less, if any, cooperation with other people. When Greeks took up settled agriculture roughly 2,000 years later than the Chinese, the farmers operated more like businessmen than peasants. They quickly moved into commercial production of wine and olive oil for trade rather than growing food for subsistence. All in all, the Greek way of life allowed individuals to act on their own to a great extent, and since there was less pressing need to maintain harmony, it also allowed for disagreement and debate. Nisbett cautions, however, that such a materialistic explanation does not imply inevitability—"just that, other things being equal, physical factors can influence to some degree economic factors and consequently cultural ones."

This is not to say that all humans everywhere don't share some basic instincts and needs or that humans are infinitely malleable, but the differences arising from culture are real and significant. Habits of thought can either help or hinder our ability to deal with our growing planetary emergency. Modern Westerners still demonstrate the same implicit assumptions about change in the world that one finds in ancient Greek philosophers: They are disinclined to believe in dramatic change and tend to assume that "future change will continue in the same direction, and at the same rate as current change." In sharp contrast, Asians today, like Taoist and Confucian philosophers, still believe the world is constantly changing and see the world as one characterized by reversals rather than change in the same direction. They are inclined to regard stability as the exception, not the rule.

The findings from this fascinating body of research left me feeling that the human enterprise could use the combined strengths of both cultures at this critical time, for each offsets the chief weakness of the other. This best-of-both-worlds merger would join our Western inclination to take action, our sense of agency, with the Asian expectation of change and holistic sensitivity to context and relationship. But to make our way in this new historical landscape, we also need to expand our sense of relationship and context beyond the human realm to embrace the physical world and the Earth system as well. The planetary scale of the human enterprise has put us in a new ecological situation—one that requires appropriate habits of mind and new kinds of perception.

In the middle of the eighteenth century, the Swedish naturalist and father of modern biological taxonomy, Carl Linnaeus, named our species *Homo sapiens*—Man the wise, an ambitious name reflecting Enlightenment optimism that, it seems, we have yet to earn. If one were to choose a less questionable and more fitting name, it would certainly be Man the flexible. In taking the path of flexibility, our human forebears responded to a changing world more and more by cultural innovation and the transmission of knowledge rather than through physical evolution. But the story is not simply that physical evolution gave way to cultural evolution, but rather that the two coevolved in a creative dialectic and together enabled our ancestors to make the great evolutionary leap that gave rise to modern humans. Thus, the mastery of fire leads to cooking, and cooking frees energy otherwise required for digestion, thereby paving the way for the physical evolution of larger brains. Similarly, the combination of biology and culture was essential to overcome the physical constraints limiting brain size and complexity. The most complex brain that ever evolved emerged by extending the period of brain development well into an individual's teenage years and by depending on culture to play a role in the developmental process. It was an ingenious solution to an evolutionary conundrum, and it further enhanced human flexibility. Just as our ancestors became untethered from any specific environment, human behavior became increasingly untethered from specific genetic programs. In this way, humans have become a single species with many natures and diverse cultures.

This cultural flexibility helped ensure our survival in the short run by allowing for swift response to new conditions and in the long run by enabling humans to live in a wide range of environments all over the world. Extinction is far less likely in species with diverse populations and extensive geographic range. For this reason, Rick Potts notes, "dividing the human endeavor into divergent ways of life raised the chance of perpetuation" in the course of extreme evolutionary trials.

Over the past 50,000 years, modern humans spread outward from Africa and outposts in the Mideast and colonized the rest of the world, moving west into Europe and east into Asia and migrating in waves to Australia, Eastern Siberia, the margins of the Pacific, Japan, and the Americas. Between 10,000 B.C. and A.D. 1500, daring pioneers ventured into the most extreme and challenging environments, eventually establishing an enduring human presence in the Arctic, on islands in the deep Pacific and the Indian Ocean, in tropical rainforests, and in great sand deserts.

While many marvel at the technological feats of the modern era and such accomplishments as space travel, I find the greatest wonder in the things that humans have achieved through great cultural sophistication and simple but effective technology. I am awed that Micronesians and Polynesians traveled the open waters of the Pacific navigating by the stars and that a species that evolved in Africa's tropics colonized the most extreme climate on Earth in the high Arctic. The ability of Inuit people to survive in this harsh, utterly unforgiving environment depended on an extraordinary knowledge of their habitat and refined and sophisticated tools: warm, efficient clothing employing different furs for different purposes, the toggle harpoon for hunting seals through the ice, the use of specialized sled dogs for transportation, hunting, and protection from polar bears, and well-designed kayaks made of bones and skin. These nomadic peoples living at the pole may not have had a complex civilization, but their culture was an impressive accomplishment.

For much of our species' history, our ancestors lived in separate groups but maintained strong social and trade connections with other groups—a strategy that helped them survive in times of local scarcity. The archaeologist Clive Gamble describes the prehistoric networks formed by human groups at the height of the last ice age as "insurance policies . . . for bad times on the mammoth steppe." Excavations at the 23,000-year-old Kostenki settlement on Russia's Don River 250 miles south of Moscow have provided a fascinating picture of how modern humans lived during this period on the cold, dry, barren steppes. In the shelter provided by a chalk ravine, these ancestors built houses by digging pits into the loose, windblown soil as far down as the permafrost, roughly three and half feet below the surface, and then constructing a vaulted roof with the gigantic tusks of the wooly mammoths that roamed the arid plains. Archaeologists assume the builders insulated the bone framework with moss and turf and perhaps covered it with animal skins, and judging by the re-creations of mammoth-bone houses at the National Science Museum in Tokyo and the American Museum of Natural History in New York, this provided a sturdy, cozy, attractive home. The dig at Kostenki has also unearthed two very large oval longhouses constructed of bones and tusks, measuring 120 by 45 feet, which had a row of hearths running down the center. Since there were few trees in this ice age landscape, mammoth bones also provided the fuel for the fires in these hearths, where the inhabitants cooked and warmed themselves. These adept householders dug pits into the permafrost—ice age freezers—to provide safe long-term storage for meat and fish. The bone and ivory needles found at the site, along with the bones of such animals as arctic foxes, suggest that those living in these bone houses trapped animals for fur and sewed warm, tailored clothing as defense against the cold. They carved and decorated bones and pieces of ivory, including figurines of brown bears, mammoths, and women with unnaturally large breasts and buttocks wearing belts, bracelets, and necklaces.

Although the Kostenki community was largely self-sufficient, excavations there and elsewhere indicate that they were part of a larger social system, a pattern of "interregional interaction" and "pancontinental alliances." Gamble believes that shared rituals, art, and tradition linked the far-flung communities in this ice age network and facilitated "the unimpeded flow of people" in this sparsely populated landscape "to contract marriages and negotiate partnerships of social and economic value." Items unearthed at Kostenki also testify to long-distance trade in essential or valued goods: flint from a source 125 miles distant and shells from the Black Sea, 375 miles away. This period, Gamble writes, was notable for "the intensity of alliances" that extended from what is now Russia into Austria, the Czech Republic, and Poland, a network vital to "long-term survival" as the ice age reached its peak. Through this web of connection, these autonomous communities scattered across the steppe could exchange important information, trade essential or desirable goods, and seek out help in hard times.

Human culture can be a paradoxical phenomenon. The very cultural talents that ensured human survival during extreme evolutionary trials have also allowed us to become, in Potts's words, "a manufacturer of crisis" and an increasing threat to the survival of this complex, global civilization and perhaps even to our species. Culture, which arose as a vehicle for flexible behavior, can often give rise to dangerous inflexibility, because cultures involve self-perpetuating mechanisms to maintain their integrity across generations. A culture can change dramatically, as Western civilization has, and at the same time stay surprisingly the same in fundamental ways through millennia. If Confucius and Aristotle were to return today, each could surely recognize whether he belonged in Beijing or in Athens, and not simply because of the physical appearance of the residents. Cultural inertia can lead to stagnation and block effective responses to new conditions, or, once the forces are set in motion, culture can propel deep change with breathtaking speed.

With growing anxiety about our failure to combat global warming, many have cited the demise of the Viking colony on Greenland as a cautionary tale about destructive cultural attitudes and unwillingness to adapt to changing conditions. In the late tenth century, during an exceptionally benign climatic spell for Europe known as the Medieval Warm Period, the Norse outlaw Eric the Red set out with a fleet of twenty-five ships bearing settlers to Greenland. Fourteen ships completed the perilous voyage, and the newcomers first established a settlement in a grassy area in the southwestern part of the island and later another to the west. The colony, which prospered for several centuries and grew to a population of perhaps five thousand at its peak, made its living from dairy and sheep farming supplemented by some hunting and carried on a brisk trade in luxury goods with Norway. In exchange for such Greenland products as ivory from narwhal and walrus tusks, polar bear skins, and live gyrfalcons, the Norse Greenlanders acquired iron, timber, tools, and other necessities as well as such finer things as wine, church bells, and stained-glass windows. But in time, with worsening weather and changing trade patterns, the trade with Norway dwindled and then stopped altogether. The Greenland colony became isolated, and by the middle of the fourteenth century, as the Medieval Warm Period gave way to the cold of the Little Ice Age, it died out.

Archaeologists have long debated the reasons for its mysterious disappearance, but many now favor the theory, advanced by archaeologist Thomas McGovern of the City University of New York, that this colony on the far western edge of the medieval frontier was the victim both of a worsening climate and of inflexibility in the face of new environments and changing conditions. The Greenland ice cores have provided an invaluable record of climate conditions in the period leading up to the colony's demise, which has allowed McGovern and other archaeologists to better interpret what has been found in decades of painstaking excavation. Even though they endured for four hundred years, the Greenland settlements were tenuous from the start because the colonists tried to transplant a Scandinavian style of farming—based on pigs and cows—to a place that had a very short growing season, far more fragile soils, and an unstable climate. "They lived on the edge from the beginning to the end," says paleoecologist Paul Buckland from the University of Sheffield in the UK, and they finally abandoned one of the two major settlements after two decades of cool summers. Even in good years, the growing season was so limited that the Norse had difficulty growing and storing enough hay to feed their farm animals through the winter. The greater mystery perhaps is why the Norse stubbornly pursued this unsuitable, unsustainable way of life for four centuries, making them so vulnerable that "it didn't take much to push them over."

The archaeological evidence shows that the settlers remained stubbornly wedded to their customary Norse ways, although the deepening cold made it more and more difficult to continue to farm. They continued to wear European clothes; women dressed in low-cut woolen gowns, even though the sealskin parkas and fur pants of their Inuit neighbors were better suited to the increasingly frigid climate. Nor did they adopt the Inuits' hunting gear, such as the toggle harpoon, which would have allowed them to hunt seals through holes in the ice during the critical time of winter scarcity. The starving colonists of Greenland's western settlement consumed their cattle and killed and ate their dogs even though, McGovern says, "there were seals in the fjord, right under the ice." Alternatives were immediately at hand but were ignored, McGovern believes, because of cultural bias and fear of losing their European identity. Culture played a key role in the colony's extinction.

"Any people not willing to reconsider old ideas as they step into new contexts may be doomed to live in a fatal cultural dead end," warns anthropologist Paul Bohannan, who has examined the way cultural traditions can become "cultural traps." When faced with a change in their physical or social environment, all cultures face the danger of persisting in maladaptive patterns that can lead inexorably to collapse. It has happened repeatedly and famously through human history, according to Bohannon, including to the ancient Greeks, who organized themselves into city-states but could not take the next step and figure out ways for them to cooperate and coexist. During the twenty-seven-year-long Peloponnesian War, the fifth-century-B.C. Greek city-states, which celebrated warrior heroes and warfare, engaged in a series of conflicts that impoverished their people, destroyed cities, and brought the golden age of Greece to an end. In the aftermath, the weakened, contentious Greeks became vulnerable to invasion and were later conquered by the Macedonians. Bohannon warns, "As conditions change, any religious explanations or political convictions that stifle thought and preclude questioning become deadly. When that happens, the very culture that helped its people to solve whatever problem they in fact solved becomes a trap that destroys everything their ancestors worked for." At such times of change, people must be willing to examine the fundamental unspoken premises that underlie their culture and beware of continuing to promote cultural values that are no longer fit for the world they are living in.

Cultures usually contain conflicting ideas that can help them evolve over time to meet changing circumstances. By shifting emphasis or retrieving long-overshadowed parts of its own thought, a culture can alter deep assumptions without losing all coherence or self-destructing. In the Christian tradition, for example, the humility and democratic fellowship that St. Francis of Assisi extended to his "sisters" the birds and his "brother" the wolf of Gubbio persisted as a minority perspective along with the majority belief that God had given humans dominion over Nature. Western culture has altered its assumptions about Nature several times during the past fifteen centuries.

At the moment, however, our current civilization—which originated in the West but has come to dominate the world through the global spread of its economic system, technology, and values—seems caught in such a cultural trap; it is drawing us deeper and deeper into trouble. We exhibit an alarming unwillingness to admit that the world has changed and to recognize the dangerous folly of business as usual. Our way of life has been unsustainable for some time, and the world is unraveling around us, yet we are loath to question the goals and values of the modern cultural experiment that put us into ever greater jeopardy. There is reason to despair at our failure of imagination, at our inability to conceive that this modern culture is not the only or best way of being human. There is reason to fear that we might share the fate of the ancient Greeks, who did not seem even to know they were in a cultural bind and so did nothing to avert the approaching collapse. We are in the midst of a cultural crisis as well as a planetary emergency. If we fail to recognize this and to do our utmost to escape from the obsolete ideas and practices that underlie our culture, our civilization surely won't survive. It is profoundly at odds with the world we now inhabit. We can face the future head-on and find creative ways to adjust and redesign our civilization, or we can resist until physical circumstances force this unavoidable change upon us—perhaps brutally.

Hope is a precious resource. Our evolutionary legacy argues that we have the inherent capacity—flexibility, imagination, and creativity—to meet the challenges ahead and avoid a cultural dead end. In a world prone to abrupt changes and extremes, simply staying alive has been no small achievement. The fossil record bears witness to the evolutionary ordeal and to the sobering fact that our closest hominin relatives did not survive. The historical record tells us that cultures perish when they become trapped in self-destructive syndromes, though cultural evolution is more forgiving of failure than biological evolution. Classical Greek civilization may have vanished, but the Greek people lived on through the calamities of collapse and conquest and still inhabit the land of Aristotle. We should not forget, through the worst that lies ahead, that a long line of resilient survivors stands behind us. Humans have done this before. This is a true and honest hope with deep roots in the story of our endurance within a volatile and capricious nature.

6

Playing Prospero:

The Temptations of Technofix

If by your art, my dearest father, you have
Put the wild waters in this roar, allay them.
—MIRANDA, IMPLORING THE MAGICIAN PROSPERO,
IN *THE TEMPEST*

Cultural traps. The words echo ominously. The ancient Greeks could not relinquish their celebration of warrior heroes or an inclination to destroy one another. The Norse Greenlanders clung fiercely to their European identity and ways, even when it had become mad to do so. To what do we cling? What prevents us from coming to grips with the predicament of this era? What is our own madness?

The modern dream of controlling nature dies hard. Indeed, as scientist and philosopher René Dubos observed, "the urge to control nature is probably the most characteristic aspect of Western civilization." We continue to operate with outdated assumptions about the nature of Nature and human power, ideas that are now demonstrably obsolete. We fashion forecasts and smooth-curved scenarios for unpredictable, nonlinear natural systems that have already proved capable of worse than the worst case we have imagined. We command vast knowledge, but fail to recognize that our ignorance is far more important in this new historical era.

There is a dangerous, perhaps fatal, mismatch between this long-standing dream of control and the unique perils of the planetary era. Nothing illustrates this clash more vividly than the flood of proposals to "solve" global warming. As the reality of accelerating climate change becomes inescapable, some scientists and engineers have remained confident that there must be a quicker and easier solution to global warming than giving up fossil fuels. Or, in some cases, they believe that the time has come to consider desperate measures. The best hope, they contend, is for some sort of "silver bullet," a technology that can stop the storm. Like Shakespeare's sorcerer Prospero, they have faith that the same modern magic that set the climate tempest in motion can also bring it to an end.

This line of thinking has given rise to proposals for geoengineering—*deliberate* planetary-scale manipulation of the Earth's metabolism to counter the unintentional disruption caused by humans. Schemes to offset global warming include wild-sounding plans to mount mirrors in space to deflect sunlight, to blast sulfate particles into the stratosphere to act as a sunshade, and to distribute iron or nitrogen across stretches of ocean to stimulate the growth of tiny floating plants that pull carbon dioxide out of the atmosphere.

A technological fix is the quintessential modern response. The great appeal of geoengineering is that it promises we can escape this dilemma without disrupting the status quo, without making fundamental changes in our energy system or in the global economy. But looking for technological solutions—whether bold geoengineering or more modest energy alternatives—is a piecemeal approach that focuses on individual symptoms of this far broader human crisis. It tends to simplify the world and how we perceive what ails us. Thus, the many aspects of global change affecting Earth's metabolism get reduced to a climate problem, and that in turn is reduced to a problem of carbon dioxide and fossil fuels, when other human activities and other greenhouse gases also play important roles. So the "solution" is either alternative energy or geoengineering to offset the problems caused by fossil fuels. Focusing narrowly and simplifying, as is the modern wont, short-circuits thinking about the systemic nature of our dilemma. If the cause of the problem is simply "fossil fuels," then one never gets to the larger question about the scale and nature of the modern human enterprise, which is problematic regardless of its energy supply. Worst of all, according to Will Steffen, a noted authority on global change, problem solving that looks first and foremost to technological fixes "allows us not to think about our relationship to the planet."

The recent surge of geoengineering proposals into mainstream scientific discussion gives the impression that it is a new avenue of exploration, but the idea has been hovering on the margins of the climate debate for as long as increasing levels of carbon dioxide have been a worry. Planetary manipulation was, in fact, the first and only solution proposed to President Lyndon Johnson in 1965 when climate change began to register as a possible threat. Johnson's Science Advisory Committee counseled that "deleterious" climate change might be solved through such countermeasures as spreading buoyant, shiny particles on the ocean surface to reflect more of the sun's radiation back into space. There is not even a mention of a more reliable policy—limiting the use of fossil fuels.

Growing environmental consciousness over the past forty years has, however, fostered resistance to this modern impulse to manipulate nature rather than accommodate it. In the face of skepticism and even revulsion to the idea, advocates of geoengineering sometimes avoid the term and speak in euphemisms that sound benign and bureaucratic. Thus, proposals for planetary-scale manipulation advance under such banners as "carbon management" or "radiation management." Whatever the words, modern habits of mind still reign in eco-nomic and political circles, so it seems to be far easier for many lead-ers to entertain truly radical proposals to purposefully alter the entire Earth system than to rethink our goals and redesign the way we live.

In truth, humans have been manipulating planetary systems for some time in an unprecedented and risky global experiment, as sci-entists Roger Revelle and Hans Suess of the Scripps Institution of Oceanography pointed out a half century ago. Revelle, who was among the first to caution about the consequences of rising levels of carbon dioxide, played a key role in establishing a climate observatory on the flank of the world's largest volcano, Mauna Loa in Hawaii. In 1958, a young investigator named Charles Keeling began continuous precise measurements there, which show the inexorable year-by-year rise of at-mospheric carbon dioxide. Now, more than fifty years later, this series, known as "the Keeling curve," is considered one of the most important scientific records in the study of global change. Before Keeling's data showed otherwise, most scientists believed the oceans would easily ab-sorb the additional carbon dioxide released by burning fossil fuels; few worried that it would accumulate in the atmosphere or warm the planet. As this great global experiment progressed, we changed the planet's at-mosphere significantly, not just with carbon dioxide but also with other chemical by-products of modern life, without understanding what the consequences would be. With the appearance of the ozone hole and the growing danger of runaway climate change, we are at last finding out.

Some fear that just this mere promise of a technological fix could erode the will to do the difficult and necessary thing: reduce green-house gases. Given that the emissions rate has been rising rather than falling, this is a significant worry. At the same time, without a variety of technologies, intelligently directed and deployed, it will be impossible to strategically retreat from our destructive engagement with essential planetary systems. The question isn't one of the use of technology or not, but what kind of technology and at what risk. And to what end?

Until recently, this planetary experiment had been an unwitting one. Now, however, those who advocate planetary-scale geoengineering propose that we make the manipulation of Earth systems a deliberate undertaking. This makes geoengineering something far more than another technological solution to an environmental problem. It would mark a conscious decision that humans would presume to take on the active management of the ancient metabolism that has sustained Earthly life for billions of years. The pioneer in planetary physiology, James Lovelock, recognizes that geoengineering could buy us time, but he warns that setting humanity on this course could lead in the long term to "the ultimate form of slavery." The more we meddle, the more we assume responsibility for keeping the Earth a fit place to live. He compares this to deciding to manage the complex automatic processes of our own bodily systems. This is a vastly greater leap than the giant leap that made humans a planetary force in the twentieth century. The implications are daunting.

Practical questions spring immediately to mind, as well as profound ethical and philosophical questions, which demand serious reflection. First, is it possible? Perhaps we do not understand the Earth system well enough. Perhaps the risk is too great. Who should decide to take such a risk, and how should they go about it? Can anyone predict the consequences if we miscalculate and something goes badly wrong?

Beyond whether it is wise to deliberately manipulate this Earth system, is it *right* to do so? It may be the case that such deliberate manipulation would be justified in some circumstances but not in others. In recent decades, we have come to understand a great deal about the history and nature of the Earth and our origins within it—knowledge that bears on the question of the appropriate relationship between humans and this living system of which we are but a part. We must consider our place within this commonwealth of life. Do we have any duties to it? And what obligation do people living today have to the generations that will follow us? If we were to choose to manipulate and "manage" planetary systems instead of cutting human sources of greenhouse gases, what enduring burden would we bequeath to them and what risk would that pose to long-term human survival? The ethical traditions we have inherited from the modern era—which focus on individuals and rights and on utilitarian measures of human happiness—are as inadequate for addressing these moral questions as the era's belief in a passive, mechanical nature is useless in investigating climate change.

The new view of a collaborative, interrelated Earth has inescapable moral implications. If the Earth is not just a vast storehouse of resources for the human enterprise, but rather a single living system, an entity whose metabolism makes all planetary life, including our own, possible, then this fact challenges the narrow humanism of contemporary thought. How is it that our Western tradition cannot, as the moral philosopher Mary Midgley points out, find intrinsic value in anything except the human individual or apply the words *sacred* or *sanctity* to anything save human life?

For some, including the distinguished physician and science essayist Lewis Thomas, the picture taken of Earth from the moon left little doubt that it was a living whole that inspired reverence and wonder. "Viewed from the distance of the moon, the astonishing thing about the earth, catching the breath, is that it is alive. The photographs show the dry, pounded surface of the moon in the foreground, dead as an old bone. Aloft, floating free beneath the gleaming membrane of bright blue sky, is the rising earth, the only exuberant thing in this part of the cosmos. . . . It has the organized, self-contained look of a live creature, full of information, marvelously skilled in handling the sun." It is absurd to insist upon the sanctity of humans while denying the sanctity of this larger life that enfolds us, of the overarching process that gives Earth its green vitality and has done so for a longer time than the human mind can conceive. Is there not sacredness as well in the living Earth?

The notion of a geoengineering "fix" fits comfortably with the modern image of nature as a machine. Questions of right or wrong arise readily, however, when the experiment involves a living Earth and the great self-regulating metabolism upon which everything beneath Earth's bright blue sky depends.

Geoengineering comes with a questionable pedigree and a colorful history of grandiose dreams, half-baked and preposterous schemes, and advocates who sometimes fit the stereotype of the mad scientist. All this has earned geoengineering the reputation of "fringe entertainment," as climate scientist Gavin Schmidt put it, rather than serious science.

In the quarter century following World War II, the belief that humanity's conquest of nature was within reach spurred U.S. and Russian scientists alike to chase the dream of controlling the weather or even permanently altering the climate. Various research projects during this period aimed to increase rainfall, to improve the harsh Russian climate by using soot or nuclear devices to melt the Arctic ice cap, and to add climate and weather alteration to the arsenal in warfare. An advisory committee in the late 1950s reported to President Dwight Eisenhower that "weather modification could become a more important weapon than the atom bomb." During the Vietnam War, the United States ventured secretly into climatological warfare with extensive cloud-seeding operations over the Ho Chi Minh Trail, the maze of footpaths, dirt roads, and rivers used to supply the North Vietnamese army. Whether Project Popeye, which flew more than 2,600 cloud-seeding missions, actually succeeded in increasing rains and bogging down the enemy supply line in mud is unclear, but it certainly did not turn the tide of the war in favor of the United States. When news of this weather warfare came to light, it set off controversy and protest that culminated in 1977 in an international treaty banning the hostile use of "environmental modification techniques."

Although the U.S. and Soviet governments invested millions in programs to modify and control the weather, these costly research efforts never yielded convincing results. Even worse, in one dramatic case, disaster followed rainmaking efforts, suggesting that scientists had about as much control as the legendary sorcerer's apprentice, who, in the tale by the German poet Goethe, incited spirits he could not command. In June 1972, torrential rains poured down on the Black Hills of South Dakota after a federal and state–sponsored cloud-seeding operation, delivering fifteen inches in six hours. This deluge, close to the area's total *annual* rainfall, unleashed floodwaters that roared through the community of Rapid City, claiming 238 lives, injuring more than 3,000, and devastating everything in its path. Although federal funding for weather modification plummeted in the mid-1970s, the temptation to keep trying has been irresistible. Cloud seeding continues in many countries, and with worsening drought in the western United States, interest in weather modification is again on the rise in Congress. During the Beijing Olympics in August 2008, Chinese meteorologists fired more than a thousand rockets containing silver iodide and dry ice into approaching rain clouds that threatened to dampen the opening ceremony. Afterward, officials from the city's weather-modification bureau claimed that this effort had successfully diverted the clouds and ensured dry skies for the spectacle.

Perhaps the most remarkable expression of the unbridled hubris that prevailed in the mid-twentieth century appeared in a scientific manifesto titled "Man Versus Climate," published by Soviet Russian scientists N. Rusin and L. Flit in 1960. Citing the difficulties presented to Russians by Arctic ice, frozen stretches of permafrost, dust storms, deserts, and water shortages, they argue, "If we want to improve our planet and make it more suitable for life, we must alter its climate." Their triumphant survey concludes with several sentences that distill the utopian fantasies that René Dubos called the modern era's "dreams of reason":

We have described those mysteries of nature already penetrated by science, the daring projects put forward for transforming our planet, and the fantastic dreams to be realized in the future. Today we are merely on the threshold of the conquest of nature. But, if on turning the last page, the reader is convinced that man can really be the master of the planet and that the future is in his hands, then the authors will consider that they have fulfilled their purpose.

In the United States, no one embodied these reckless and extravagant ambitions more vividly than geoengineering's most enthusiastic advocate, Edward Teller, who is often described as the father of the hydrogen bomb and has been long rumored to be in part the inspiration for film director Stanley Kubrick's portrait of modern madness, the nuclear scientist Dr. Strangelove. Over the years, Teller championed U.S. nuclear superiority to the Soviet Union, arguing that such an arms race helped ensure world stability. At the same time, he generated a long list of mad-sounding schemes involving the "peaceful" use of nuclear bombs. His memorable brainstorms included proposals to use hydrogen bombs to dig a deepwater harbor in Alaska for coal and oil transport, to excavate a second canal across Panama, to extract oil from the tar sands in Alberta, and even to "modify the weather." A recent biography concludes that Teller may have been crazy like a fox: There is evidence that his "peaceful atom" civilian projects were a cover for military activities—a ploy to continue weapons testing in the event that international talks then under way ended in a test ban.

All these projects proved unwise, unsafe, and infeasible, but Teller nevertheless wielded substantial influence with U.S. presidents and Congress throughout the Cold War. As a member of the White House Science Council, he helped convince President Reagan of the need for a space-based missile defense system using lasers and

satellites that was popularly known as Star Wars. Other leading nuclear scientists mounted a campaign against the proposal, arguing not
only that it was a bad idea on several counts but also that it simply
would not work. Teller, however, was undaunted. His faith in human
ingenuity and technological prowess was so extreme that he ignored
or simply dismissed risks and hazards. His response to growing concern about radioactive particles raining down worldwide from nuclear
weapons testing was typical: Fallout was "not worth worrying about."

In 1997, when he was approaching ninety, Teller waded into the
climate debate with a paper on geoengineering titled "Global Warming and Ice Ages," which argues that humans have the technological
means to counteract global warming or prevent the onset of the next
ice age. The paper makes the case that it makes no sense to control
carbon dioxide emissions when it would be far cheaper and easier to
solve global warming by injecting particles into the stratosphere to
block some of the incoming sunlight. The cost for this technofix, Teller
and his coauthors calculated, would be, at most, 1 percent of the estimated $100 billion a year it would take to reduce emissions. Again
Teller paid little attention to the possibility something might go
wrong, making the cure worse than the original disease. It was all quite
simple, he asserted in an opinion piece in the *Wall Street Journal*: "Injecting sunlight-scattering particles in the stratosphere appears to be a
promising approach. Why not do that?" A decade later, the editors of
the influential financial magazine the *Economist* paid tribute to Teller's
early enthusiasm, posting an online story about geoengineering and
"how big science can fix climate change" under the ironic headline
"Dr. Strangelove Saves the Earth." Until his death in 2003, Teller remained a caricature of America's over-the-top technological optimism. He was a true believer to the very end in a shiny faith, one that
had begun to corrode in the 1960s as concern mounted about escalating environmental damage and technology's dark side.

* * *

The first Earth Day in 1970 and the burgeoning of environmental concern had brought a sea change in public attitudes. Utopian schemes to "improve" the planet for human benefit ran headlong into the growing conviction that it is unwise to interfere with nature's complex systems. Nevertheless, a discussion of geoengineering continued quietly in scientific circles but focused on a new mission. Instead of fantasizing about the benefits of an ice-free Arctic, some researchers now began to consider how they might use geoengineering to counteract global warming and maintain the existing climate. Teller was not the first to suggest putting particles into the stratosphere to reflect away sunlight and cool the planet. In 1977, Russia's leading climatologist, Mikhail Budyko, had put forward the idea of "artificial volcanoes," so called because the proposed remedy mimicked the explosive delivery of sulfate particles to the stratosphere in volcanic eruptions. Budyko, however, was far more tentative than Teller. He judged that climate science would have to advance significantly before scientists could predict how this might affect regional weather and that until then, it would be unwise to tinker on a global scale.

Most climate scientists have been even more wary and ambivalent than Budyko about resorting to planetary-scale manipulation to solve global warming, a reluctance evident in the long series of scientific reports on the climate problem over more than four decades. After the initial report to President Johnson in 1965, no U.S. scientific panel expressed the same unqualified optimism about an easy technological fix for global warming. By the early 1970s, the pendulum of opinion had swung to the opposite extreme: Two reports on climate change from this time were deeply skeptical about the prospects for geoengineered solutions. Nevertheless, scientific panels convened by the National Academy of Sciences to assess the climate problem presented reports in 1983 and 1992 that considered the possible use of geoengineering to counter warming.

As climate change became an increasingly pressing policy issue in the 1990s, however, talk of geoengineering virtually stopped cold, in large part because climate scientists judged it was better to press for sure solutions—quick and deep cuts in fossil fuel emissions—than to raise the tenuous hope of a technofix through risky, unproven planetary manipulation. Most of the recent scientific assessments have, therefore, made little if any mention of radical solutions that would aim to offset one human intrusion into planetary systems with another. As climate scientist Wally Broecker noted early in the climate debate, purposeful climate modification has long been "one of the few subjects considered taboo in the realm of scientific inquiry (ranking perhaps number two after inquiries into possible racially related differences in intelligence)."

Then, in the summer of 2006, the subject of geoengineering came roaring out of the scientific closet. Dismayed by "grossly unsuccessful" political efforts to reduce emissions, Nobel laureate Paul Crutzen reluctantly decided he had to confront the hitherto unthinkable—the prospect that the world may have to resort to geoengineering as a desperate last-ditch measure to brake a rapid warming. In a paper published in the journal *Climatic Change*, Crutzen urged serious investigation of proposals to offset rising temperatures by injecting sulfate into the stratosphere in much the way Budyko, Teller, and others had suggested. In strictly scientific terms, the paper contained nothing that was really new, but, primarily because of who was now making the case, it had a stunning impact. Unlike Teller, Crutzen is no wild-eyed technological true believer, but rather a sober, widely respected atmospheric scientist and one who had delivered an early warning about how human activity might damage the ozone layer. In the apt description of Stanford University climate scientist Stephen Schneider, Crutzen is an "environmental worrier."

Crutzen's latest worry is that inaction is increasing the risk of rapid warming or large surprises. The best solution, as Crutzen stresses repeatedly in his paper, is to reduce carbon dioxide emissions, but this appears to be, as he put it, "a pious wish." The world needs a 60 to 80 percent cut to stabilize levels in the atmosphere, but in the past decade, emissions have not been declining, but rather rising faster than ever before. Since world leaders signed the Rio Convention on Climate Change in 1992, global emissions have climbed from 6.1 billion metric tons of carbon a year to 8.5 billion in 2007. If climate change arrives in leaps, it will require a quick response, so the only option, as he sees it, will be the problematic remedy of geoengineering. Once the technology is in place, Crutzen argues, it would be possible to inject reflective particles into the stratosphere on short notice and offset any rapid temperature rise within six months.

There is another worry that fuels Crutzen's sense of urgency. Air pollution has so far spared us the full brunt of the warming. For all the problems it creates, burning fossil fuels has also had one surprising benefit: The soot, ash, sulfur dioxide, sulfate, and other airborne particles it creates have been screening out solar radiation. At the same time, however, these pollutants have been causing more than 500,000 premature deaths around the world every year, so governments initiated clean air programs to combat this health threat. As a result, these cooling airborne particles have declined by 2.7 percent a year. Reducing this type of pollution will *boost* global temperatures, but there is great uncertainty about how much. One recent modeling study concluded that even with conservative assumptions, the loss of this pollution sunscreen could push temperatures way beyond the upper limit currently projected by the IPCC. "Pessimistic climate scenarios are much more plausible than had been thought," said the study's lead author, Meinrat Andreae, an atmospheric researcher and colleague of Crutzen's at the Max Planck Institute of Chemistry in Mainz, Germany. "If our model is right, things could become totally uncontrollable in the second half of the century." In Crutzen's mind, this argues for a backup plan, "an escape route" to save us from outright catastrophe. Andreae, on the other hand, is adamantly opposed to geoengineering and reportedly urged his friend Crutzen to refrain from publishing this paper. His protest is moral rather than scientific: "It's like a junkie figuring out new ways of stealing from his children."

Ralph Cicerone, president of the National Academy of Sciences and an eminent atmospheric scientist, has also been deeply involved in propelling the topic of geoengineering from the fringe into the scientific mainstream. In an editorial appearing in the same issue of *Climatic Change* as Crutzen's paper, Cicerone acknowledged that geoengineering does not enjoy broad support and that leading scientists had opposed publication of the paper. But like Crutzen, he sees the need for an insurance policy in the event of a full-blown emergency, insurance in the form of a reputable research program that can winnow the merely risky geoengineering schemes from the manifestly mad. In the meantime, Cicerone has suggested that scientific leaders consider a moratorium on any large-scale geoengineering experiments and give careful thought to such matters as establishing a body to oversee such interventions.

Geoengineering is a complex topic in that it involves the question of goals as well as the merits and risks of any proposed technology for altering the planetary metabolism. Though no one these days is as brash as the utopian Russians who declared that humans can "master the planet," the idea of active management of planetary systems for maximum human benefit has not disappeared. In a provocative essay titled "Earth Systems Engineering and Management," Brad Allenby, an Arizona State University professor of civil and environmental engineering, made the case that humans have no choice but geoengineering.

Like Bill McKibben in his early work on climate change, Allenby argues that pervasive human intrusion has brought about the end of nature—a conclusion based on a modern view of the world that sets humans and nature in stark opposition. In a world of such dichotomy, any human influence on natural systems renders them "artificial," so Allenby asserts the Earth is "a human artifact" and contends that humans now have an ethical responsibility "for engineering and managing the carbon cycle, and the climate system as a whole." In his view, "the most important ethical question we as a species will have to answer over the next decades" is "what kind of world do we want to design?" Allenby seems to harbor no doubt that the future is *entirely* in human hands, writing boldly, "Difficult as it is to accept—there is no natural history anymore: there is only human history. What these systems will be in the future is a human decision, a human choice. Having realized this, we cannot escape the ethical responsibility for that choice." On the practical matter of climate change, his essay questioned the current focus on limiting the use of fossil fuels and instead favored looking for a technological fix that would make it possible to burn fossil fuels without fueling climate change.

The climate scientists now considering the possibility of geoengineering have more modest goals. Most do not see geoengineering as an alternative to cuts in greenhouse gases, nor is their aim to remake the world according to human design. Rather, they are looking for a way to keep the current climate system from going completely haywire. Some, like Crutzen and Cicerone, see the need for a backup plan that can go into action on short notice and save the world from runaway warming or abrupt shifts in the climate system. Another prominent climate researcher, Thomas Wigley of the National Center for Atmospheric Research in Boulder, believes that geoengineering could buy time for the difficult transition from fossil fuels. In an article in *Science*, he makes the case for a strategy that combines geoengineering—a stratospheric sulfate sunscreen along the lines Crutzen proposed—with cuts in carbon dioxide emissions, arguing that this approach has distinct advantages over either alternative alone. A modest geoengineering program would allow for more gradual cuts and thus reduce the cost and technological difficulty; more ambitious geoengineering efforts could perhaps save us from consequences—soaring temperatures and rising seas in coming decades—that we can no longer hope to avoid through emissions reductions alone.

There is also the possibility, noted in an editorial in the journal *Nature Reports: Climate Change*, "that not a single one of the myriad schemes would actually work in reality." These are proposals, not proven technologies. "To rely on geoengineering as a solution to climate change would be massively irresponsible, akin to using gambling as a way to get out of debt."

Others have more fundamental objections. One of the strongly dissenting voices is Will Steffen of the Australian National University, a leading researcher on global change who formerly headed the International Geosphere-Biosphere Program that studies this planetary problem. In Steffen's view, the root problem with schemes to deliberately modify Earth's systems to counteract global change is this: Technofixes rely on a modern linear logic ill suited to the planetary era. "Such geo-engineering approaches are exceptionally dangerous," he warns. "They are based on simple cause-effect logic and ignore the fact that global change is a complex, interactive phenomenon acting on a single interlinked planetary system. Unintended consequences of geo-engineering approaches are highly likely, are very difficult to predict, and could well lead to problems as severe as those they were intended to solve."

The various geoengineering schemes to remedy the warming caused by CO_2 fall under three broad categories: blocking the sun, boosting land and ocean processes that naturally absorb carbon dioxide, and capturing and sequestering carbon dioxide so it does not accumulate in the atmosphere. These approaches all have serious shortcomings, and as a general rule, the low-cost solutions are also the most questionable.

The sunshade proposals aim to intercept and reflect sunlight in low-altitude marine clouds, higher up in the stratosphere, or farther out in space. Climate modelers estimate that any project to offset warming caused by a doubling of carbon dioxide in the atmosphere would have to deflect roughly 1.8 percent of incoming sunlight.

Since Budyko first suggested the idea of creating "artificial volcanoes," scientists have had a chance to observe the worldwide cooling following one of the largest natural eruptions of the century. In mid-June 1991, the Philippine volcano Pinatubo blasted 20 million tons of sulfur dioxide as high as twenty miles into the stratosphere during a spectacular nine-hour eruption. This immense cloud of gas produced airborne sulfate particles that stayed aloft for three years and reflected sunlight back to space, cooling the Earth's surface at the peak of its shading by as much as 1.3 degrees F. At the same time, however, these chemical changes in the stratosphere added to the ozone destruction caused by CFCs and led to a record loss of ozone.

Advocates of the human-made volcano solution point out that it is surprisingly inexpensive, can be quickly deployed, and can be discontinued in the event of unexpected consequences. Those brainstorming about how to deliver reflective particles to the stratosphere have come up with all kinds of suggestions, ranging from balloons, artillery guns, and a vast fleet of high-altitude airplanes to a man-made "mini-volcano" on an equatorial mountaintop that would create a hot plume to transport sulfur gases that form reflective particles into the stratosphere. A more fanciful idea proposes blowing the gases aloft through a thin fifteen-mile-long pipe carried upward by kites. Since the particles eventually make their way back to the surface of the Earth, whatever the delivery method, they would have to be replenished every two to three years to maintain the sunshade, and if carbon emissions keep climbing, this sunshade effort will have to deliver ever larger quantities to the stratosphere to offset the warming. So, like the Red Queen in *Alice in Wonderland*, we would need to run faster and faster simply to stay in place.

Others have been proposing ways to intercept and deflect sunlight in space before it ever reaches the Earth's atmosphere, perhaps by deploying a fine mesh parasol or 55,000 orbiting mirrors. At the annual meeting of the National Academy of Sciences in 2006, Roger Angel, a professor of astronomy and optical sciences at the University of Arizona, presented his concept for a sunshade that would be mounted at a spot between the Earth and the sun some 932,000 miles up, in a region where the gravitational fields of the sun and the Earth cancel each other out. He proposes to launch 16 trillion transparent, wafer-thin, sun-refracting discs measuring about three feet across and weighing a gram. According to his calculations, it would take ten years to deploy this number if twenty launchers shot off packets containing 800,000 discs every five minutes. The project is daunting to contemplate and would cost a few trillion dollars over twenty-five years—expensive, but only half of one percent of world GDP. But it has advantages over artificial-volcano schemes, for it would not further alter the composition of the atmosphere, could be fine-tuned using solar-powered radio receivers, and could remain in place for perhaps half a century. On the other hand, given the magnitude and expense, it might be better to invest the same resources into changing the energy system.

Researchers at the University of Edinburgh are at work on a more down-to-earth strategy that would boost the cooling capacity of clouds. This scheme, first proposed in 1990 by meteorologist John Latham of the National Center for Atmospheric Research, aims to double the number of reflective droplets in the low, lumpy layer of stratocumulus clouds that forms over parts of the world's oceans, on the theory that these brighter clouds would reflect more sunlight back to space and thereby cool the Earth. The researchers believe they can accomplish this by spraying saltwater into the air; winds will loft sea salt and other particles upward and thereby stimulate the formation of more cloud droplets. As currently proposed, a fleet of unmanned, wind-powered sailing vessels equipped with satellite navigation would sail back and forth, dragging large propellers that would act as turbines and generate the power needed to create the spray.

Regardless of the technology, the sunshade strategy simply treats one symptom of carbon dioxide—heat; it does not *solve* anything. The excess carbon dioxide humans release causes other kinds of significant environmental disruption besides warming, and chief among these is the growing acidity of the oceans. As some of this excess carbon dioxide is taken up by seawater, it forms carbonic acid, which dissolves coral reefs and seashells. Shell-forming plankton at the bottom of the food web in the Southern Ocean—foraminifera—are already having difficulty making shells to protect themselves from predators. Such plankton troubles are bound to cascade through the ecosystem that depends on them, affecting krill, seals, whales, and the fish and shellfish harvested for human food. If the oceans continue to grow more acidic, marine life will be devastated.

Nor will sunshade geoengineering necessarily maintain the climate system and current weather patterns. One big concern is that climate change will shift rainfall patterns and make already dry places even drier. If geoengineers create a sunshade in the stratosphere, however, drought problems will likely be even more severe. Two studies have used historical data to investigate how past volcanic eruptions affected rainfall. In both cases—Pinatubo in 1991 and the Icelandic volcano Laki in 1783—eruptions coincided with disrupted rainfall patterns, widespread drought, and failure of the monsoon rains that are essential for agriculture in Africa and Asia. Drought and famine hit Egypt following Laki's eruption.

A climate-modeling study on the impact of a sunshade arrived at a similarly grim forecast of decreased rainfall in all of the Americas save Canada, in all of Africa south of the Sahara, and especially in tropical areas. The reason is twofold: less evaporation from the oceans and less water moving from plants to the atmosphere. Both of these processes are part of the global water cycle that moves water around and brings rain. In the oceans, the decline in evaporation is a consequence of lower temperatures as an engineered sunshade cancels out warming. The change in plants stems from the fact that carbon dioxide levels would keep rising despite the sunshade. Plants pump huge quantities of moisture from the soil, which they release into the air through pores in their leaves—moisture that contributes to clouds and rain. In a high–carbon dioxide environment, plants do not have to open these leaf pores as wide for photosynthesis, so less water escapes to the atmosphere. This all adds up to dry skies and less rain.

By far the biggest hitch in the artificial volcano strategy is that it requires an unflagging commitment of centuries: five hundred years, or, if we do not make major emissions cuts, even as long as a millennium. Since it will take five centuries, even under the best scenario, for the oceans to remove 80 percent of the excess carbon dioxide and return atmospheric concentrations to a tolerable level, the artificial sunshade will have to stay in place long after humanity's flamboyant era of fossil energy ends. And, if anything should interrupt this geoengineering effort, the world would quickly confront a doomsday scenario. Without the stratospheric haze or sunscreen technologies to offset the consequence of extraordinary levels of atmospheric CO_2, temperatures would suddenly soar upward at a rate *twenty times faster* than they are rising today, causing unimaginable havoc in human and natural systems and with it, the real danger of human extinction. This institutional challenge is without question a far greater obstacle than any technological difficulties. "We would need centuries of trusted global climate controllers," Stephen Schneider points out. "They would have to operate continuously, reliably, and by consensus. My objection to geoengineering is not that we couldn't figure out a way to do it—though there are side effects we can't anticipate. I'm more worried about centuries of trusted controllers."

It is hard to imagine that anyone with even a passing knowledge of the vicissitudes of human history over the past millennium would think this long-term commitment could be a prudent gamble.

Another family of geoengineering schemes seeks to tackle the emissions problem squarely by accelerating the natural processes that remove carbon dioxide from the atmosphere. Only half of the carbon dioxide released from fossil fuels has accumulated in the atmosphere; the rest has been soaked up by the world's oceans or by trees and plants on land. In the oceans, floating organisms called phytoplankton pull carbon dioxide out of the atmosphere for photosynthesis, so increasing their populations by fertilizing stretches of ocean could help draw down more of the excess carbon dioxide and in theory slow global warming. Current proposals would add iron or nitrogen to ocean waters, depending on which nutrient is in short supply. Since the concept was first put forward in 1990, a dozen experiments around the world have demonstrated that seeding ocean water with iron will, in fact, cause one of the explosive increases in plankton numbers known as blooms.

Recently, however, interest in ocean fertilization has expanded beyond the scientific quest for a possible climate remedy. With the value of the global carbon market more than doubling in 2007 to $65 billion, some Silicon Valley technocrat entrepreneurs are chasing the hot business potential of ocean fertilization. In this market—which includes both those subject to mandated reductions under the Kyoto climate treaty as well as voluntary participants seeking to reduce their "carbon footprint"—businesses and individuals can "offset" their emissions by investing in carbon-reduction schemes, such as windmills, or regulated companies can install technology to reduce or eliminate carbon dioxide emissions and then sell emissions allowances they no longer need. The cost of offsetting a ton of carbon dioxide through ocean fertilization—estimated at between $4 and $15 a metric ton—would be a fraction of that for constructing alternative energy sources or for capturing and sequestering the carbon dioxide from a coal power plant, which by comparison is estimated to run about $50 a metric ton.

In 2007, two private companies—Planktos, based south of San Francisco, and Ocean Nourishment Corporation, in Australia—were already touting the benefits and pursuing large-scale trials in bids to grab a piece of the carbon-offset market. Russ George, the head of Planktos, boasted that the company "restores ecosystems and slows climate change." He described his controversial plan to dump tons of iron near the Galapagos Islands off Ecuador and elsewhere in the ocean as a "voyage of recovery," a "remediation" effort to restore levels of plankton, which he claimed have been falling abnormally. The Planktos website directed visitors to a carbon store where it advertised that you could "erase your carbon footprint" by buying offsets for airplane trips and other activities that contribute to global warming. For a mere $10, Planktos promised that a traveler can "negate" the climate consequences of a round-trip between New York and Los Angeles. The head of the Australian effort, Ian Jones, who also heads the Ocean Technology Group at the University of Sydney, claimed that his plans for adding nitrogen to ocean regions lacking this nutrient would not only pull more carbon dioxide from the atmosphere, but also enrich the food web and increase fish stocks needed to feed growing human numbers.

Another new company in the race to commercialize ocean fertilization is San Francisco–based Climos, headed by Silicon Valley entrepreneur Dan Whalley, who made a fortune during the dot-com boom with a travel-reservations website. Whalley has hired his mother, Margaret S. Leinen, a prominent researcher on ocean sediments and climate history who is a former official at the National Science Foundation, as the company's chief science officer. He has enlisted other prominent figures—including former presidents of the American Association for the Advancement of Science and of the Woods Hole Oceanographic Institution—for his panel of science advisers.

Leading ocean and climate scientists remain skeptical of ocean fertilization schemes, warning that it may not do much to remedy global warming and may well create new problems. Climate scientist Wally Broecker, who has studied factors that limit how much iron fertilization can reduce atmosperic levels of carbon dioxide, judges that "oceans fertilization is not going to make a significant difference" in solving the CO_2 problem. Others, including Sallie Chisholm, an MIT researcher in biological oceanography, warn that ongoing ocean fertilization could actually have the perverse effect of *producing* greenhouse gases. By encouraging the growth of oxygen-consuming organisms, such operations would likely reduce or deplete oxygen levels in the deep ocean, she explained, making conditions there hospitable to organisms that release methane and nitrous oxide, which are more powerful greenhouse gases than carbon dioxide. Molecule per molecule, the heat-trapping power of nitrous oxide is three hundred times greater than CO_2. These schemes could also wreak havoc with the ecosystems in tropical waters, which are adapted to a low level of nutrients. "Thousands of species depend on this ecosystem," Chisholm noted. "When you fertilise it, you disrupt all that, just as you do when fertiliser runs off the land into streams and causes 'dead zones' in coastal water." Ian Jones, the leading proponent of nitrogen fertilization, counters that the amount of nitrogen won't be enough to create "dead zones," but acknowledges that the process will inevitably change ecosystems.

Because of such concerns, opposition has been growing. In June 2007, the intergovernmental scientific committee of the London Convention on ocean dumping issued a statement of concern about the rush to commercialize ocean fertilization, warning that "knowledge about the effectiveness and potential environmental impacts . . . is insufficient to justify large-scale operations." The committee recommended that the parties to the convention consider whether such operations are legal under the treaty and, if so, how they should be regulated. In February 2008, Planktos canceled its first planned field experiment because its investors had pulled their support from the company, and later sold its research vessel to pay off bills. At the same time, however, Climos, which had successfully attracted venture capital, was seeking permits to conduct its own experiment by spreading iron over roughly four thousand square miles of ocean. But in May 2008, the 191 nations who are party to the United Nations Convention on Biological Diversity called for a moratorium on large-scale commercial ocean fertilization operations, citing the absence of reliable data regarding potential risks.

Entrepreneurs are looking as well to opportunities on land. Planktos had also been involved in projects that seek to boost the amount of carbon dioxide soaked up and stored in trees by expanding woodlands. With a flair for publicity, Planktos's Hungarian subsidiary, KlimaFa, made a gift in 2007 to the Vatican of the carbon-cancelling benefits from a tree-planting project along the Tisza River in Hungary. In theory, the thirty-seven-acre forest, which has been named the Vatican Climate Forest, will make the Vatican the world's first carbon-neutral state—provided that it will in fact absorb as much carbon dioxide each year as the Vatican produces. Klimafa's gambit was good enough to attract the attention of the *New York Times*, which ran a story about the gift under the headline "Vatican Penance: Forgive Us Our Carbon Output." Several European governments and the computer maker Dell have invested in carbon offsets to be provided by trees that KlimaFa promises to plant to restore degraded Hungarian land.

There is no question that plankton and trees absorb carbon dioxide as they photosynthesize and grow, but fostering plankton blooms and planting trees won't make a dent in global warming unless the carbon is locked away and does not return to the atmosphere. In the case of plankton, the benefit to the atmosphere depends on how much of the organic matter from plankton blooms sinks to the deep oceans and stays there. Given the complexity of doing research in the challenging environment of deep oceans, answering this question is no easy matter. Based on field research, scientists estimate that *only a few percent* of each plankton bloom makes it to the deep ocean. Predators such as zooplankton and microbes gobble up the rest, recycling the carbon in the plankton back into carbon dioxide that can move into the atmosphere. Even making it to the deep ocean is no guarantee that the carbon will stay out of circulation in the long term. Modeling studies estimate that one century after a month of continuous ocean fertilization, more than half of the quantity of carbon sinking to the depths, perhaps even as much as 98 percent, will have been recycled and back in circulation rather than safely sequestered. This means that almost all of carbon dioxide "erased" through the purchase of an ocean fertilization offset could wind up back into the atmosphere.

The same question applies to forests, especially at a time when climate change will, according to forecasts, shift growing zones, increase disease and pests, and bring drought and more frequent forest fires. Trees are already burning at a record rate around the world. As three wildfires burned in Southern California in November 2008, Governor Arnold Schwarzenegger described the growing fire hazard in a TV interview. "We used to have fire seasons only in the fall. But now the fire season starts in February already," he explained. "Through global warming, we have now a fire season all year round." The 2008 California fire toll, 1.4 million acres, fell just short of the record loss in 2007 of 1.5 million acres, the greatest devastation in at least four decades. In the western United States, the number of yearly large, devastating wildfires since 1986 has jumped fourfold, and the extent of the area burned is up sixfold compared to blazes between 1970 and 1986.

Catastrophic fires have also plagued other regions over the past decade. In the summer of 2007, wildfires swept across a half-million acres of old-growth forest and scrublands in Greece, killing sixty people and hundreds of thousands of farm animals. In central Siberia,

15,000 square miles of forest went up in smoke in 2003. In Alaska, vast areas of forest are succumbing to drought, insects, and fire. In 2004 alone, 6.6 million acres burned. Portugal, Spain, the Balkans, and other parts of southern Europe have also suffered devastating fires in recent years, and fire disasters are bound to increase, because the Mediterranean region faces drought and desertification in a warming world. What are the chances that trees planted to "erase" the carbon dioxide burden from a cross-country air flight will be alive and well and still storing carbon a half century hence? If the current fire trends continue, forests may become a *source* of carbon dioxide and a *cause* of global warming rather than the solution touted by climate entrepreneurs.

"Burying" global warming by sequestering carbon dioxide is the most mainstream of the proposed technological fixes. The idea is to capture carbon dioxide from burning fossil fuels before it escapes into the atmosphere, compress it, and then find a way to store it securely. Scientists have focused first on coal-fired power plants, which now account for 38 percent of the world's CO_2 emissions, and, with a thousand new coal plants on the drawing boards worldwide, their contribution to the climate problem will possibly double in coming decades. By the end of the century, energy analysts forecast that coal will account for half of the world's energy supply. "The climate problem," says Daniel Schrag, who directs Harvard University's Center for the Environment, "is a coal problem."

Researchers say the basic technology, often referred to as "clean coal," already exists to burn coal without releasing carbon dioxide, but most U.S. power companies have shown little interest in carbon capture and storage because, depending on the method, it could increase the cost of electricity by as much 90 percent. The biggest emitter of carbon dioxide in the U.S. power industry, American Electric Power, is the notable exception. It is in the process of planning and building two new coal plants, in Ohio and West Virginia, designed to be ready for carbon capture and storage technology—a move that will make it far easier and cheaper to retrofit the plant if limits are placed on CO_2 emissions. AEP is aiming to complete the first of these in 2010.

The method used to capture the carbon dioxide depends on the design of the coal plant. In the traditional coal-fired steam power plant, the gases from combustion would go into an absorption tower rather than up a smokestack. There chemicals would absorb the CO_2 and proceed to a stripper tower, where the chemical mixture would be heated to release the concentrated CO_2. There had been high hopes that a new way of burning coal, the integrated gas combined cycle, would make carbon capture far less expensive, but the FutureGen project to demonstrate the new method's feasibility on a commercial scale stalled in 2008, when the Bush adminstration scrapped this public-private venture because of cost overruns. (The Obama administration may, however, resurrect it.) Nor has there been significant progress elsewhere, because governments and the private sector have been similarly reluctant to come up with the necessary funds. For all the talk of carbon capture, there is only one power plant in the world retrofitted to do so, a gas-fired boiler at Lacq in the south of France that went into operation in April 2009. The 60,000 tons of carbon dioxide captured each year will be stored nearby in a depleted gas field.

Climate campaigners in the United States, meanwhile, are demanding that carbon capture and storage begin immediately and that no new plant be built unless it is ready for retrofit. Without technology to capture the carbon dioxide, a new 1,000-megawatt coal plant releases 6 million tons of carbon dioxide a year and will do so for its life span of half a century. Only a dozen of the 150 coal-fired power plants planned in recent years in the United States have been designed for easy retrofit. By the end of 2007, however, the coal rush in the United States began losing steam because of rising construction costs, anticipated climate legislation by Congress, and rising concern among utility regulators about coal's contribution to global warming. By one count, utilities abandoned plans or did not secure permits for fifty-nine plants, and fifty others face court challenges. The remaining will likely face challenges when they seek permits.

Capturing carbon dioxide at a power plant prevents more carbon dioxide from entering the atmosphere, but that doesn't solve the problem of the excess 100 parts per million that is already there and causing alarming changes. In September 2008, a leading European climate expert, John Schellnhuber, said the time has come to take seriously the idea of removing man-made emissions from the air, a plan he describes as "atmospheric restitution." He is just one of several prominent scientists warning that political leaders have been too slow to cut emissions and that time has run out. In his view, the odds are now greater than 50 percent that the future will bring dangerous levels of climate change because of the excess carbon dioxide already released. The idea of "scrubbing" the air is an ambitious goal that has the air of science fiction, but three groups of researchers are at work on industrial technology to do it. If air scrubbing proved feasible and affordable on a grand scale, and this is a daunting *if*, this would be the only geoengineering scheme that would offer an assured way to restore the atmosphere to levels normal before the Industrial Revolution.

Air capture of carbon dioxide is not as hard as one might imagine, and it would not require visionary undeveloped technology, according to David Keith, a leading expert on geoengineering and part of a University of Calgary research team. Researchers there are taking a conservative approach and building on technology we already have to pull carbon dioxide molecules out of the air and condense them into a pure stream of carbon dioxide that can be buried. All the chemicals involved are safe and well understood, Keith notes, because Kraft paper mills—which make the brown paper used in grocery bags—have been using a similar chemical process for years.

In April 2007, Global Research Technologies, a research and development company based in Tucson, Arizona, working with Klaus

Lackner from Columbia University, announced that they had achieved the first successful demonstration of a prototype device for grabbing carbon from the air. In the GRT design, the full-scale air scrubbers will be about the size of a shipping container and remove a ton of CO_2 a day. When Lackner, a physicist with Columbia's Earth Engineering Center, first presented his idea for giant carbon dioxide filters at an international meeting on coal and fuel technology in 1999, the reaction was "utter disbelief." In 2003, however, Lackner, Wally Broecker, and others met with the founder of Lands' End clothing, Gary Comer, who then put up the money that gave GRT its start and allowed the company to develop the demonstration Lackner described as "an exciting step toward making carbon capture and sequestration a viable technology."

Air capture will be expensive and, as currently envisioned, would require twice as much energy as capturing 90 percent of the CO_2 from a power plant, so some experts on carbon capture remain skeptical. Lackner estimates that it will cost $200 to capture and store a ton of carbon dioxide, making the annual global cost perhaps in the range of $5.6 trillion. But the remedy provides a way around intractable problems arising from the larger number and variety of carbon dioxide emitting sources that now make cuts extremely difficult. Air scrubbing amounts to a retrofit of the energy *system* as a whole, notes Keith; it offers an alternative strategy to retrofitting or replacing all the countless sources of CO_2 around the world—individual power plants, cars, and homes—which will be both expensive and challenging. More than eight thousand industrial facilities around the world account for 60 percent of annual CO_2 emissions. The 700 million passenger vehicles in the world contribute another 10 percent, and, according to Keith, air scrubbing might prove a cheaper solution for dealing with these emissions than the much touted hydrogen fuel-cell car. Air-scrubbing plants would have great flexibility because they would not be bound to emissions sources: They could be built anywhere, on any scale, and could mop up CO_2 emitted anywhere in the world. If this technology becomes feasible on a large scale, which is far from guaranteed, it would mean that the dangerous excess of carbon dioxide already in the atmosphere and already causing frightening changes might be brought down to safer levels.

It also offers a new way to deal with questions of fairness and equity. Countries that got rich by burning fossil fuels could take responsibility for removing their share of emissions from the atmosphere, although the absence of accurate, comparable historical data might make it difficult, if not impossible, to negotiate and execute any agreement to do so. Developed countries with 20 percent of the world's population have contributed more than three quarters of the excess carbon dioxide that has accumulated in the atmosphere since the Industrial Revolution. For the United States—which has alone contributed almost 30 percent of the 321.08 billion metric tons of excess carbon added to the atmosphere since 1850—the historical debt would be about 94 billion metric tons. The European Union, which now has twenty-seven member countries, stands in second place with roughly 27 percent of the historical carbon debt, or about 85 billion metric tons.

Once the carbon is captured at a power plant or sucked out of the air, the idea is to put it back into the Earth for secure, long-term storage. Here again, the key questions are whether the carbon dioxide will stay put and remain out of circulation and whether it will cause any other problems in the meantime. Schemes for storing CO_2 on land include injecting it into oil and gas reservoirs, coal seams, or saline aquifers on land. In the oceans, carbon storage advocates have proposed pumping it into deep ocean waters or below the floor of the ocean at depths approaching ten thousand feet. It is also possible to chemically transform CO_2 into a highly stable magnesium carbonate rock that would keep it out of the atmosphere indefinitely.

Researchers had judged that land storage in saltwater aquifers lying in sandstone 5,000 to 8,000 feet down would be the most likely choice, but a test injection of carbon dioxide at such a site in Texas raised questions about its security and safety. Monitoring showed that the carbon dioxide made the salty water as acid as vinegar, so it began dissolving the surrounding rock that kept it in place. If acidic water opens pores and fractures in the rock, the carbon dioxide could escape into the atmosphere and the nasty brine could migrate into nearby aquifers and pollute water used for drinking and irrigation.

The escape of carbon dioxide from an underground storage site poses two kinds of hazards. If the gas leaks out slowly, it would defeat the purpose of capture and make the whole effort a waste of time and money. If leaking carbon dioxide were to accumulate in pockets beneath the ground and then escape in a rapid burst, it could kill people, who would be suffocated as the carbon dioxide displaced oxygen. Sudden releases of carbon dioxide near volcanoes and seismic faults have taken a considerable toll. In 1986, 1,700 people in Cameroon died after 1.2 million tons of carbon dioxide exploded out of the depths of Lake Nyos, which lies in a volcanic crater. The gas release killed all living things within a fifteen-mile radius. A blast of carbon dioxide from a volcano in Indonesia suffocated 142 people. If carbon dioxide stored underground escaped and seeped into a basement, it would be odorless and invisible, but it would kill anyone who entered. There is also concern that injecting large amounts of carbon dioxide underground could trigger earthquakes. Because there would always be some danger that carbon dioxide would find a way to escape, any carbon dioxide storage sites on land would require long-term monitoring.

A scheme for ocean disposal by injecting carbon dioxide into the depths proved controversial for a variety of reasons, including worries that it would damage marine life and that ocean currents would mix the carbon dioxide and allow a good part of it to return to the atmosphere. To prevent such a return, scientists say it would be necessary to inject CO_2 at depths greater than 11,500 feet. Initial experiments with deep ocean injection have shown that liquid CO_2 released at such depths reacts with seawater to form a slushy clathrate compound, which is denser than seawater or CO_2 in liquid form. More recently, scientists at Harvard, MIT, and Columbia have made a case for injecting CO_2 beneath mud at the bottom of the ocean. Although the cost will be about 25 percent higher than shallower burial, the researchers say this method will provide leakproof, permanent storage for all the carbon emissions the United States could ever generate. While the carbon dioxide stored underground on land is buoyant and prone to escape, at the high pressures and low temperatures in deep-sea sediments at 9,800 feet below the surface, the gas becomes liquid and denser than surrounding seawater. At great pressure, the pool of stored CO_2 will dissolve slowly into the surrounding ocean over millions of years.

Perhaps the most fail-safe way to lock up carbon dioxide indefinitely is to turn it to stone through a chemical reaction with such widely available minerals as serpentine and olivine. This process of mineral carbonization, which occurs in nature, creates magnesium carbonate, which is similar to limestone. Researchers at Arizona State University are working to make this a feasible way to sequester carbon. The possible obstacles are high energy demands for the chemical process and high cost, as well as the physical fact that this method would consume huge quantities of rock to store millions of tons of carbon.

None of these technofixes, however, will give us the power of Prospero to stop the storm. Even the most promising of these possibilities, scrubbing carbon dioxide from the air, is far from a silver bullet. This technology, if it proves out, would dramatically alter prospects by making the already dangerous and still-growing concentrations of carbon dioxide in the atmosphere *reversible*. But bringing carbon dioxide down to safe levels will not return the climate system to its previous state and save the world we've known. The most miraculous of technological breakthroughs cannot undo the damage that has already been done. The planetary system has been seriously perturbed and great changes are already under way on many fronts. In the century ahead and beyond, the Earth's response will continue to play out, no doubt in surprising ways.

However inviting the prospects shimmering on the technological horizon, geoengineering "solutions" and the promise of a technofix down the road lead us easily into temptation. It can encourage neglect of what can be done *now*, because possible solutions ahead seem better or easier. Former President George W. Bush often used future technology as an excuse for inaction, touting research on hydrogen fuel-cell "freedom cars" while rejecting proposals to improve the efficiency of today's vehicles. One energy economist quipped, the freedom car "is really about Bush's freedom to do nothing about cars today."

The large geoengineering plans, such as sunscreens or air scrubbers, offer another type of temptation: planetary-scale technological intervention as "insurance" in case emergency suddenly escalates into catastrophe. It is hard to object to a backup plan, especially as the world has not yet halted the rise of emissions, much less embarked on the deep reductions that are required. Insurance, however, often has a perverse effect: It can undermine rather than increase security. The promise that someone or something will be there to bail you out if the worst happens encourages imprudent behavior. The number of mountain rescues has increased now that hikers carry cell phones. The National Flood Insurance Program available to people living in coastal communities aimed to discourage development in high-risk areas, but it has instead increased the risk of death and disaster by stimulating development in these danger zones. Similarly, geoengineering fosters the notion that technology can rescue us from climate hell, if it comes to that, and thereby discourages early, prudent action to head off the worst danger. Banking on speculative technologies that may not pan out is an irresponsible and immoral gamble that may leave our children and grandchildren with only bad choices, or no choice at all.

Perhaps worst of all from a moral perspective, the pursuit of geo-engineering offers a way to pass the problem on to the next generation under the guise of virtuousness. "Climate change," says University of Washington philosopher Stephen M. Gardiner, "is a perfect moral storm." When we direct funds to research programs instead of undertaking meaningful, immediate action to cut emissions, we may convince ourselves that we are doing something, planning ahead, and thereby salve our consciences, when in truth we are dodging our obligations to those who preceded us and those who will follow. Those left reaping the whirlwind in the decades ahead may curse the moral collapse of their ancestors and their capacity for self-delusion.

The political hazards of deliberate planetary manipulation are as formidable as the moral pitfalls. The technologies that scientists and engineers regard as "insurance" to safeguard the human future may precipitate dangerous new kinds of international conflict and the possibility of a planetary-era arms race in geoengineering technology.

If geoengineering becomes the chosen response, the obvious question is, Who is going to make decisions that are truly global in scope, and how? Who, if anyone, will be approving, overseeing, and policing any use of geoengineering? If the time comes when the Earth needs a sunshade, there must be a guarantee that, once started, it will continue for centuries. If the monsoon fails following some geoengineering effort, there must be some authority to mediate the dispute about what caused it or compensate those who claim damages. As Stephen Schneider once suggested, such claims are inevitable, so it would be unwise to do this without some plan for "no-fault climate disaster insurance" to provide compensation. And how is it going to be possible to distinguish plain old bad weather from climatological warfare? In a geoengineered world, a catastrophic hurricane or a devastating drought can generate suspicion, paranoia, and conflict.

The problems of the planetary era clearly require some manner of global governance, but our first attempts at this have failed miserably. Gus Speth, dean of the Yale School of Forestry and Environmental Studies and an early leader on global problems, describes the current state of affairs bluntly: "The climate convention is not protecting the climate, the biodiversity convention is not protecting biodiversity, the desertification convention is not preventing desertification, and even the older and stronger Convention on the Law of the Sea is not protecting fisheries. Nor are they poised to do so in the immediate future."

The planetary system binds us more tightly in a common destiny than the economic system. No one will be secure in a world with runaway warming. Yet governments that willingly concede some of their sovereignty to promote economic expansion will not do the same to protect planetary systems.

The United States has a particularly shameful record in this regard. Long the world's largest contributor to global warming, the nation refused to take part in the Kyoto protocol and even tried to obstruct the efforts of the 178 nations that ratified the climate treaty. U.S. leaders may still entertain the delusion that America is a moral leader and a light to the world, but future historians will no doubt see it as a Nero among nations, fiddling callously as the crisis has escalated. The thing future generations may remember about this new world power is its profound moral failure at this critical time. Without a bold and radical change of policy, the United States is headed for an infamy that will be remembered like Nero's for millennia.

Of course, the United States has not been the only obstacle to greater global governance and effective treaties. It is not clear what, short of a terrifying climate jolt, would propel nations jealous of their sovereignty to finally take the unifying step that is needed to manage global problems effectively. In the planetary era, it is not enough to simply *think* globally while acting locally. We need bold, coordinated global action to address global-scale change.

In the absence of some means to arrive at a collective decision and to provide oversight, all sorts of conflicts and tensions are almost inevitable. What happens if a single country decides to opt for some sort of planetary manipulation instead of reducing its emissions? What if other countries object that the project is too risky? If it becomes possible to scrub carbon dioxide from the air and reduce carbon dioxide levels, the question of who gets to choose the climate could spark serious disputes. For the moment, leading figures in Russia are inclined to think that global warming won't be half bad. After all, since the Soviet era, the country's scientists have been trying to figure out a way to alter the climate and take the bite out of the brutal Russian winter that defeated both Napoléon and Hitler. One recent head of the Russian environmental agency, Konstantin Pulikovsky, is sanguine about climate change: "For our great northern country, I don't today see any imminent problems for the next 100 years at least." During his time in office, President Vladimir Putin commented lightheartedly about the advantages, saying, "We'll spend less on fur coats."

Russia has also been at the head of the race to claim rights to the land beneath the no-longer-icebound Arctic Ocean, which may contain oil, gas, and minerals. In August 2007, a Russian submarine crew journeyed to the North Pole and planted a flag on the ocean floor. Given the ambition to exploit Arctic resources and an upbeat assessment about the benefits of climate change, the Russians are likely to balk at a plan to reduce carbon dioxide to 280 to 300 parts per million— a target that would return carbon dioxide levels to the safe range for the climate system. "You can imagine some kind of arms race of geoengineering where one country is trying to cool the planet and another is trying to warm the planet," said Ken Caldeira of the Carnegie Institution at Stanford University.

Could a climate standoff come to dominate international politics in the twenty-first century in the way the Cold War dominated the second half of the twentieth century? Could capitalist leaders decide to act on their own to stabilize temperatures because governments have failed? One can easily imagine the dozen richest people in the world deciding to step in and deploy vast numbers of air scrubbers because the deteriorating climate is threatening the economy.

The ultimate hazard in pursuing a technological fix for climate change, however, is that it only addresses one symptom. Technology may push back immediate constraints, but it does not solve the problem of ongoing exponential economic growth and the scale of the human enterprise, which must be faced sooner or later. Analysts forecast that the global economy will expand by another 500 percent in this century. Even with the greenest energy technology and geoengineering, one has to ask whether this is possible on a planet already in a state of emergency. The contemporary celebration of economic growth always brings to mind a warning from the nineteenth-century British art and social critic John Ruskin: "That which seems to be wealth may in verity be only the gilded index of far-reaching ruin." Technology may allow us to avoid for a while longer the hard systemic questions about how to live productively within the imperatives of a finite Earth, but this avoidance leaves those who will finally confront this challenge with ever worsening options.

In seeming despair, the economist Nicholas Georgescu-Roegen asked, "Will mankind listen to any program that implies a constriction of its addiction to . . . comfort? Perhaps the destiny of man is to have a short, but fiery, exciting and extravagant life?"

I for one don't believe in destiny or in spectacular self-destruction as an inevitable human fate. While humans do not have "control" of Nature like the magician Prospero, we do have great powers of choice, if we are willing to exercise them, over the human activities that generate planetary jeopardy. At a fateful turning point in the journey of the ancient Israelites in the wilderness, the Old Testament prophet Moses described the situation starkly, saying, "I have set before you life and death, the blessing and the curse: therefore choose life that you and your descendants may live." Now at another critical moment in the longer human journey, this is our choice as well.

On Vulnerability and Survivability

Amid the danger and uncertainty of the planetary era, how does one "choose life"?

Choosing life begins with courage, the courage to confront the complexity and contingency of this world and let go of the modern illusion that we can bring it under human control. The absence of control is itself a terrifying thought, especially for anyone who has been nurtured on the dreams and promises of our current civilization. A sober look at the radical uncertainty of the human future, moreover, gives reason for real fear, the kind of primal fear that drives to the bone. But fear can be, must be, faced down rather than repressed or denied. The times are too dangerous to do otherwise. Though life for now continues with a sense of normality, the current order is no longer viable and hasn't been for some time. The deep change that lies ahead threatens to shake the foundations of natural systems and human societies alike. Courage, as Martin Luther King Jr. observed, is "the power of mind to overcome fear." It requires "the exercise of a creative will" to challenge "the forces that threaten to negate life." Fear is an emotion; courage is a mental discipline long counted as one of the supreme human virtues.

When the gravity of the planetary crisis begins to dawn on people, the common response is a frantic demand for "solutions" or "a plan" that will make things right again. Averting the planetary emergency and avoiding change, however, is not possible: We have crossed thresholds and cannot go back. The failure to confront these hard facts has itself become an added problem. In a commentary in the journal *Nature*, three scientists, who headed the IPCC's review of future impacts from climate change, warned about the air of unreality that persists even as the forecasts have grown darker. "A curious optimism—the belief that we can find a way to fully avoid all the serious threats—pervades the political arenas of the G8 and the UN climate meetings," they wrote. "This is false optimism and it is obscuring reality."

How then do we meet this changing world? What must we do to avoid the worst? And how can we prepare ourselves to weather the change that is now unavoidable? Answering these difficult questions is how we choose life in this new historical landscape.

A strategy of survivability should be aimed at enhancing security and improving the odds for those who will live in a future that may unfold in ways we can hardly imagine. First and foremost, this strategy must reorient human efforts toward two overriding goals: reducing our activities that disrupt the Earth's metabolism and, equally important, reversing trends in human systems that make us vulnerable to the potentially catastrophic consequences of this disruption. These goals are the foundation for survival in the planetary era. Even if they seem unachievable at the moment and the path forward is not entirely clear, it is vital to know where one should be headed and to embrace these goals with steady, unflagging commitment.

Reducing the human burden on planetary systems is not only necessary, it is an achievable goal that does not require continuing economic growth, as is often argued, or depend on the promise of some futuristic technology. Vaclav Smil, a leading historian and analyst on energy, makes a compelling case that it is possible to relieve the pressure on Earth's essential systems resulting from energy demands while maintaining a reasonable quality of life for humans. The questions of what Earth requires to sustain life and what humans really need to live decently are seldom asked, he notes, because they run counter to "the reigning ethos of economic growth." Though Smil's analysis is extensive and complicated, his final assessment is stunning in its simplicity: "I do not think I exaggerate when I see this primarily as an issue of attitude rather than a distinct painful choice." The challenge of narrowing the unconscionable gap between the rich and poor worlds and reducing the energy use in the affluent world that already overburdens Earth systems is, Smil concludes, primarily "a moral issue, not a technical or economic matter." True innovation will not lie in finding better answers within the current framework, but rather in asking different questions.

The obligation to endure is everyone's responsibility. The *we* I refer to throughout this book embraces all those alive today. Governments, markets, business entrepreneurs, science, technology, laws, treaties, and policy innovations all have important roles, but they cannot save humankind if each of us and the communities we belong to do not rise up and choose life by courageously facing the gravity of our situation and redirecting human efforts toward these new goals. Those who have wealth and power, especially the 500 million who count among the extravagantly rich and account for half of global carbon dioxide emissions, have the greatest burden. A group of these prominent people could provide invaluable cultural leadership in this material era by abandoning the indulgences of the wealthy—private jets, multiple homes around the world, exotic holidays, lavish consumption—in recognition that these indeed are "the index of far-reaching ruin." They could lead by moderation and demonstrated concern for the future as well as by any foundation and charitable activities they may now pursue. In doing so, their legacy to their children and grandchildren and the world at large would be far greater than the wealth they have accumulated. This wealth won't matter if current trends continue.

The defining feature of the decades ahead will almost certainly be discontinuities and surprises. The alarming thing is how unprepared we are for such a time.

* * *

Conventional wisdom holds that global warming will be primarily a problem for the world's poor, the "most vulnerable." Those living in developed countries, the world's comparatively rich, policy experts have said, will have the means to adapt to a changing world. For many who are not among the poor, this may offer comforting assurance amid news of accelerating disruption, but beware of its unspoken assumption—that climate change will proceed in a more or less manageable way. If abrupt change is the *norm* rather than the exception, it is wishful thinking to believe that modern industrial civilization can take it all in stride. Conventional wisdom will prove conventional folly if the climate changes by leaps and surprises. In the face of such shocks, the rich—who are most dependent on the goods produced by the global economy, ready access to fossil fuels, and sophisticated technologies—may find it difficult, if not impossible, to adapt. Think of what would happen if a breakdown of the global infrastructure leads to a loss of energy for even a few months. In short order, it would be far better to know how to start a fire without a match than to own a state-of-the-art Viking stove. The world's rich countries are vulnerable, more vulnerable than their political and economic leaders can allow themselves to imagine, but in ways different from the poor. Who turns out to be most vulnerable in the coming century will depend both on how climate changes and on how governments, social and economic institutions, and communities confront uncertainty in a much more volatile world.

The chance of a future punctuated by shocks and surprises grows greater day by day. From a narrow economic perspective, the explosive growth in the global economy over the past decade has been a positive development because it has provided a better living for many people in countries like China and India and expanded global wealth overall. But in the broader view, this rapid expansion of the human enterprise is ominous. In the next twenty years, current forecasts project this gargantuan global economy is going to need 55 percent more energy—energy that is expected to come mostly from coal and oil. If this proves to be the case, even with maximum improvements in efficiency, annual carbon dioxide emissions will still rise by at least 25 percent by 2030. Propelled by such an unprecedented boom, the world will hurtle faster and further into the zone of dangerous climate change—the point where the rising burden of greenhouse gases in the Earth's atmosphere begins to unleash such catastrophic consequences as the disintegration of the Greenland and West Antarctic ice sheets. Stanford climate scientist Stephen Schneider offered a stark summary of where we now find ourselves: "If the object is to avoid dangerous climate change . . . , we are twenty-five years too late. . . . The object now is to avoid *really* dangerous climate change."

The pressure on human societies from other kinds of problems is mounting as well. In the economic sphere, one sees the instability of the global economic system and the growing gap between rich and poor. Many nations struggle with population growth and the rise of megacities, as well as with recurring food crises as global food reserves have dwindled alarmingly. Competition for increasingly scarce oil has led to destabilizing price shocks. And around the world, the degradation and depletion of soils, forests, fisheries, and other natural systems continues. Thomas Homer-Dixon, a political scientist at the University of Toronto and a leading expert on social crisis and breakdown, describes these as "tectonic stresses . . . accumulating deep underneath the surface of our societies." The convergence of destabilizing forces is particularly "treacherous" and, he warns, the possibility of abrupt breakdown in social and economic systems is rising. "September 11, 2001, won't be the last time we walk out of our cities."

The menacing storm is one of our own making. We haven't, however, recognized the other half of our dilemma: that this civilization is making us ever more vulnerable to the instability and disruption it has set in motion. While industrial civilization has succeeded famously in raising living standards over the past two centuries, at the same time it has been compromising much of the adaptability that characterizes our species. The central fact about this highly specialized social and economic system is that it depends *on existing conditions.* The modern way of life is "fully predicated upon stable climate, cheap energy and water, and rapid population and economic growth," as environmental historian J. R. McNeill observes—circumstances that can be only temporary on a finite, changeable Earth. Over the past century, many societies around the world have committed without reservation to this single, specialized, fossil fuel–based strategy. In this respect, the human enterprise now has much in common with the extinct "lawnmower" species of the African savanna, which adapted superbly to one set of conditions and were extremely successful—until conditions shifted. For most of the human career, as McNeill points out, we have shared far more with rats: another species of nimble, flexible generalists and remarkable survivors.

Our inherited ideas make it difficult to conceive that our powerful civilization is in reality profoundly vulnerable. In the modern narrative, the story of civilization, which emerged some six thousand years ago, is a story in which humans progressively transcend the limits and vicissitudes of nature. The march of civilization, so we have been taught, has enhanced our security. But as the understanding of our past has grown, a far more complex picture has been emerging. Civilization has been not so much a triumph as a trade-off, and like our physical evolution, its history does not record how human societies managed to escape Nature but rather how they responded to the shifting moods of Earth. This revised story illuminates the fundamental vulnerability of the current order and the way the modern era compounded the gamble that our ancestors made on civilization. Vulnerability and survivability are two faces of the same question.

Without a recognition of this vulnerability, there is little chance of making headway on measures that may help human communities weather the coming decades.

A dark paradox lies at the heart of the six-thousand-year trend toward increasingly complex societies: The pursuit of security has set us up in the long run for catastrophe. It may be true that the cultural innovation of agriculture increased the margin of safety initially, but in the analysis of archaeologist and anthropologist Brian Fagan, the long-term consequences have been another matter. "In our effort to cushion ourselves against smaller, more frequent climate stresses, we have continuously made ourselves more vulnerable to rarer but larger catastrophes. The whole course of civilization . . . may be seen as a process of trading up on the scale of vulnerability."

In his provocative book *The Long Summer* and his other work, Fagan challenges the smug notion that rich countries have built a protective buffer in modern agriculture, industry, and shipping and will readily adapt to changing conditions. For all our technological prowess, we are arguably far more vulnerable to climate change than our Stone Age ancestors who managed to thrive in brutal and erratic climatic conditions during the last ice age. His narrative of increasing human vulnerability traces the shift from the hunter-gatherer life that humans pursued for most of our history to permanent villages, the rise of agriculture, the first cities, and onward to the twentieth-century surge of population growth, urbanization, and the global spread of the Industrial Revolution. Each step in this history, he argues, came at a considerable price—lost flexibility and fewer survival options. For our hunter-gatherer ancestors, "survival depended on mobility and opportunism, on a flexibility of daily existence that allowed people to roll with the climatic punches—by moving away, hiving off families into new territories, or falling back on a cushion of less desirable foods. Around 10,000 B.C., when farming began, anchoring permanent villages to their fields, the options afforded by mobility began to close."

* * *

Since anthropological theories have generally celebrated the progress of civilization, it was long taken for granted that our ancestors made the shift to agriculture—which happened independently in several different regions of the world—because it improved their lives. The benefits, it was typically assumed, were not only a more secure food supply but one that was better in quality and easier to acquire. Given this view, it was simply assumed that life had been punishing in the Paleolithic before agriculture and that hunter-gatherers led a nasty, brutish existence, exhausting themselves in a constant quest for food and teetering perpetually on the brink of starvation. Accordingly, the transition to agriculture, which textbooks have celebrated as the Neolithic Great Leap Forward, "liberated" our ancestors from this lot, and humans made a momentous advance in the march to civilization.

The study of ancient bones, however, tells quite a different story, which paleopathologists have retrieved in much the way forensic anthropologists help solve crimes on TV programs such as *Bones* or *CSI*. Their findings confront us with provocative questions. The Great Leap was apparently not so great in many respects, for hunter-gatherers, who did not commonly suffer from malnutrition or starvation, often lived *better* than their descendants who turned to agriculture. Taken as a whole, in fact, the evidence from most parts of the world indicates that diets became poorer as these ancestors settled down and became farmers. Even worse, it appears they had to work harder in this new way of life.

The lives of hunter-gatherers living closer to our times—recorded in historical accounts from early European contact in the seventeenth, eighteenth, and nineteenth centuries and by anthropological study in the twentieth century—lend support to this finding. Contrary to the received wisdom, hunter-gatherers generally had access to abundant, reliable, and nutritious food, and since they typically spent at most four to five hours a day securing and preparing food, they worked far less to make a living than a person today in an industrial society. Anthropologists studying the daily routine of native Australians and Bushmen in Botswana in the 1960s described a life in which spells of intermittent work punctuated leisure passed in socializing, dancing, resting, or daytime naps.

A seventeenth-century French Jesuit, Pierre Biard, who spent time with the Micmac tribe in what is now the Canadian province of Nova Scotia, provided a similar account of a rich and sufficient life. Even his refined French palate found nothing wanting in Micmac fare that shifted with the seasons from seal to moose, caribou, beaver, bear, then smelt and herring, seabirds, salmon, shellfish, and eel: "Solomon never had his mansion better ordered and provided with food." He also writes with envy about the way that life proceeded like a moveable feast as the tribe shifted from camp to camp. On the day of departure, they "start off . . . with as much pleasure as if they were going on a stroll or an excursion . . . in order to thoroughly enjoy this." With their fine canoes and great paddling skill, the Micmac could, Biard estimates, cover sixty miles a day or more in good weather, but, he writes, "we scarcely see these Savages posting along at this rate, for their days are nothing but pastime. They are never in a hurry. Quite different from us, who can never do anything without hurry and worry. . . ."

Biard wrote this around 1616, fully a century before the birth of his fellow countryman Jean-Jacques Rousseau, a philosopher who identified primitive peoples with Nature and inspired the Romantic notion of the Noble Savage. The Jesuit's account is free of this later idealization, and I take his report of the joy and gusto with which the Micmac approached life, including their good humor even during the hungry season at the end of winter, to be generally reliable. Since Biard's time, hurry and worry, it seems, have grown exponentially along with the physical scale of civilization. The now vanished way of life of the seventeenth-century Micmac provides a standard against which to measure modern madness, which devours life in multitasking mania. It reminds us of other kinds of abundance and what we may have lost on the journey described as progress.

If growing food does not ensure a better, easier living than hunting and gathering, why then did our ancestors embark upon this path? There is no simple answer to this question, and theories abound. As scientists have gained insight into past conditions through the Greenland ice core and other records, however, one thing has become clear—climate is a critical part of the story. The exceptional overall stability of climate over the past 11,700 years—the "long summer"—made settled life and farming possible for the first time in the history of our species. Whatever other factors may have played a role, this far more reliable Holocene climate was, as one archaeologist put it, the *"ultimate enabler* for early farming."

While general climatic conditions have been relatively benign in recent millennia, this period has not been uneventful from the human perspective. Extreme weather, such as catastrophic droughts lasting a century or longer, have punctuated the long stretches of favorable conditions and have altered the course of human history repeatedly. Fagan believes that these megadroughts figured prominently in the advent of agriculture in the Middle East and in the collapse of great civilizations. The threat of a new era of megadroughts looms large in the coming century, because past experience during the Medieval Warm Period from A.D. 800 to 1200, as well as projections based on climate models, suggests that even a slight warming beyond what has occurred already may be enough to push many regions—the Mediterranean, sub-Saharan Africa, the American West—over the threshold into deep and lasting drought. If this happens, it may make life difficult if not impossible for those living in affected areas. In Fagan's view, there is much to learn about how humans adapt to such climate extremes in the history of two ancient settlements in the Middle East, the long vanished Syrian village of Abu Hureyra and the Mesopotamian city of Ur, which was once the height of civilization and the most powerful city in the ancient world. In both, one sees the interplay of security and vulnerability and the prominent and decisive role that rare extreme events have played in human history.

The Levant—the sweeping arc of land at the eastern end of the Mediterranean stretching from southern Turkey through Syria, Jordan, Lebanon, and Israel to Egypt's Sinai Peninsula—became a landscape rich in new opportunities as the world warmed after the end of the last ice age. For roughly two millennia after 12,700 B.C., the region enjoyed improved rains and favorable climate. The rapid spread of oak and pistachio forests created a new habitat that provided an abundance of nourishing food throughout the year. Hunting groups, which had subsisted on game during the late ice age, shifted to a diet that relied increasingly on nuts and seeds and began settling down—a step documented by the remains of house mice, which moved in as soon as humans stopped moving around. The blessings of a benign, stable climate allowed for a new kind of hunter-gatherer life to take shape, one where people could live together in larger groups and establish permanent villages of solid round houses roofed with thatch. These people, known today as Natufians, hunted desert gazelles, harvested wild grain with flint-bladed sickles in spring and summer, and gathered acorns and pistachios in the fall.

The village of Abu Hureyra, which now lies beneath the waters of Lake Assad, created by the Tabaqah Dam in Syria, was a permanent settlement near the Euphrates River probably founded by the Natufians, where the residents carried on this kind of hunter-gatherer life for some five hundred years. Thanks to meticulous excavations before its loss to rising waters, archaeologists preserved its vital history and the record of how the villagers responded in the face of extreme climate change some 13,000 years ago. As Fagan reads this record, climatic conditions—both good and bad—shaped the evolution of this community and spurred the experimentation that led in time to farming.

No longer on the move, the people of Abu Hureyra built store-rooms and saved the surplus from good years for lean times such as a dry spell or a failed nut crop. Pistachios provided fast and easy food, because their shells split along the edges when ripe and the nuts emerged ready to eat. The acorns, however, were another matter. As at other pivotal times in the human past, food preparation played a prominent role in the way life developed. To make acorns into a palatable and nourishing food, the women spent long hours with mortars and pestles pounding the nuts and then soaking them to remove the bitter tannins. This effective processing method and the investment of more labor to do the task yielded more nourishment from small areas of land, so over time the region became crowded with the contiguous territories of permanent hunter-gatherer villages. As this happened, Fagan writes, these communities lost the social flexibility and mobility "that was as old as humanity itself" and in doing so, "crossed a threshold of environmental vulnerability." But for generations, this deeper insecurity was not apparent. As long as the rains and favorable conditions lasted, Abu Hureyra flourished and grew to a community of perhaps four hundred people.

Then, the good times ended. The cause of this reversal of climate fortune, according to the leading theory, was a shutdown of ocean currents thousands of miles away in the North Atlantic that brought on the abrupt climatic shift, discussed earlier, known as the Younger Dryas. Europe plunged back into extreme cold, the glaciers advanced again in Scandinavia, and the rains shifted in the Levant, bringing a devastating thousand-year-long drought.

Archaeological evidence shows how the community of Abu Hureyra tried to cope as the vise of drought tightened. Around 11,000 B.C., as the forest that provided fruits and nuts retreated from the area, the gathering efforts shifted increasingly to seeds of wild grasses. When asphodel, feathergrass, and other wild cereals also vanished around 10,600 B.C., the community fell back on drought-resistant clover and other less desirable foods. Then they experimented with trying to sow and cultivate the foods that were disappearing. Evidence of domesticated seeds—lentils, rye, and a type of wheat called einkorn—appears around this time, but the early foray into agriculture was not enough to sustain a permanent village. A few generations later, the only remaining option in the face of a deepening drought was moving on in search of better conditions. Around 10,000 B.C., those who were left abandoned Abu Hureyra.

It may be, Fagan suggests, that survivors sought out places with reliable water, such as natural oases, and continued to experiment with planting seeds to provide some backup for their hunting and gathering in the drought-stressed landscape. However it happened, after the warming resumed and returning rains ended the millennial drought, people eventually returned to Abu Hureyra, people who were full-fledged farmers and almost wholly dependent on cultivated grains. Five centuries later, the new settlement had grown to a town with several thousand people. The move to agriculture had advantages, but these lasted only as long as the rains and reserves. Settled life and a reliance on fewer kinds of food made the long-term risks even greater.

Agriculture opened the door to the specialized way of life we call civilization because the surplus it produced freed some part of the population from the task of acquiring food. Those who did not have to spend their days seeking sustenance could become priests, craftsmen, scribes, bureaucrats, tax collectors, and soldiers who defended the lands and crops from invaders and raiders. Increasing sophistication and growing cultural complexity did not, however, make these societies invulnerable to the vagaries of climate. Indeed, complex societies face the danger that Reggae great Jimmy Cliff warns about: "The harder they come, the harder they fall."

One of the earliest centers of civilization in world history, the ancient Mesopotamian city of Ur, is today a pile of ruins amid the desolation of the desert south of Baghdad. Its effort to respond to a worsening climate illustrates the "harder they fall" paradox that accompanies growing complexity. This once-great city near the Euphrates grew from a small Sumerian farming community that took root around 6000 B.C. during a period of ample rains and abundant floods. But with a shift in the Indian monsoon track around 3800 B.C., the winter rains faltered and the floods arrived too late for the crops. This threw the farmers into crisis, as they found themselves dependent solely on the river that now had diminished flow during the critical times in the growing season. They responded by planting barley and wheat later so the crops did not ripen before the floodwaters arrived, and over time they moved into larger cities and towns built around a more elaborate centralized irrigation system that gave greater control over the timing and quantity of water.

Ur grew into one of these cities, and with its irrigation system and the security of temple storerooms filled with grain, it flourished for more than a millennium in this difficult climate, where rains seldom brought more than eight inches a year. As Ur and the other early cities between the Tigris and Euphrates grew into complex societies, the Sumerians developed many of the fruits of civilization—the wheel, writing, math, astronomy, and written literature, including the famous Gilgamesh epic. In time, the rulers of Ur became the most powerful in the ancient world. During the city's ascendancy, one of them, named Ur-Nammu, built magnificent structures, including the Great Ziggurat—a sixty-four-foot-tall terraced pyramid once graced with hanging plants that still rises above the desert and is visible for miles on the vast flat plain.

Just a century later, the great city of Ur was staggering from two severe climatic blows that sent the region into chaos. The first was a huge volcanic eruption somewhere far to the north around 2200 B.C. that blasted vast quantities of gas and ash into the stratosphere, causing a plunge in temperature and probably diminished rains. To make matters worse, the region was just heading at this time into a 278-year drought, and according to Fagan, the moist westerlies that brought rain faltered "with catastrophic abruptness." The winter rains never came, and the annual floods of the Tigris and Euphrates failed.

As cities like Ur struggled with severe water shortages, they were at the same time facing a flood of drought refugees—nomadic herders who began streaming southward as the once fertile plains to the north dried up. The army was sent to stem this tide, and when it failed, Ur's ruler built a 112-mile-long wall named the Repeller of the Amorites. It apparently did not live up to its name, because as displaced people sought refuge, Ur's population tripled in the decades that followed, and desperation grew. Cuneiform tablets from this time record that officials in charge of the storehouses doled out grain rations by the tablespoon. Lawlessness became an increasing problem, people began leaving the city, and within a few generations, Ur collapsed. "Having arisen as a successful defense against small catastrophes," Fagan argues, "the city found itself increasingly vulnerable to larger ones." The Sumerians had become masterful at managing their environment during ordinary bad times, but when true calamity hit, their magnificent civilization fell to pieces.

Others considering the human story as a whole have also noted the civilization paradox. William H. McNeill, professor emeritus of history at the University of Chicago and one of the leading scholars of big-picture history, describes this pattern—growing mastery leading to escalating vulnerability—as the underside of the civilized condition. It is "the price we pay for being able to alter natural balances and to transform the face of the earth through collective effort and the use of tools." The classic story of civilization as "progress" belies the zigzag path humans have traveled in recent millennia. McNeill characterizes human history as "an extraordinary, dynamic equilibrium in which triumph and disaster recur perpetually on an ever-increasing scale as our skills and knowledge grow." Like Fagan, he questions the notion that this growth in social complexity has made us safer. "Modern societies, armed with contemporary levels of science and technology, are enormously powerful. No one doubts that. Yet they are no less vulnerable."

Anthropologist Joseph Tainter provides another angle on this question in his influential theory of how and why societies embark on the path of increasing complexity and often collapse in the end like Ur or, in the most famous example, like Rome. Tainter also believes that societies choose to increase their complexity in order to solve immediate problems and that this inevitably makes them more vulnerable in the long run. In contrast to Fagan, however, his analysis focuses on how the *internal dynamics* of social and political systems evolve over time and generate growing instability as societies struggle to maintain their complex structures.

Tainter's reasoning begins with the fundamental fact about both Earth systems and human ones: They require a continuous flow of energy to maintain their organization, and the greater the complexity, the greater the demand for energy. Complexity is costly, and as social and political complexity increase, a society inevitably reaches a point of rapidly diminishing returns, after which ever greater investments yield ever smaller rewards. A society then finds itself having to invest more and more just to maintain the status quo. The reason for this is simple and logical: When a society invests in complexity to solve a problem, it uses the most readily available resources first—it irrigates the land closest to the river or exploits shallow deposits of coal and oil. In time, however, it will have to move on to more difficult locations, investing more to obtain the same benefit.

Without question, modern industrial societies are already far along on the curve of "diminishing marginal returns" with regard to the fossil energy that makes their extraordinary complexity possible. Tainter sees the dilemma of diminishing returns intensifying in other aspects of the current system as well, including agriculture, mineral resources, and investment in health and education. In medicine, the easy and inexpensive cures have been won. Take penicillin, for example. It cost no more than $20,000 for the basic research leading to its discovery. Today we are investing far greater sums just to maintain the benefits of antibiotics in the face of growing drug resistance and in the effort to cure more difficult diseases.

As a society struggles to maintain the status quo in the face of diminishing returns, its vulnerability grows. This happens, Tainter writes, because it begins to lose the reserves needed to ride out periods of crisis, which he calls *stress surges*. This surge capacity might be the ability to produce more than normally required amounts in agriculture, minerals, and energy or to maintain stores of surplus from past production. Without such reserves, a society will have to respond to unexpected stress by diverting resources needed to maintain itself under ordinary circumstances, which weakens and begins to undermine the system. Once on this path, "collapse may merely require the passage of time."

Modern industrial civilization is already risking breakdown with its dependence on stable climate, cheap energy, and growth. Even worse, the current wave of globalization is upping the ante and pulling all humanity into this high-stakes game. For much of human history, trade among villages, and later among regional economies, provided rare materials and luxuries or served as a buffer against local scarcities, but the transformation of regional markets into a single global system has threatened this traditional security. Some might argue that globalization is not all that new because intercontinental trade in coffee was already under way in the thirteenth century, and a few centuries later, trading ships loaded with cocoa, sugar, tea, gold, silver, spices, furs, and tobacco were plying the oceans. But even the burst of globalization from 1870 to 1914—which emerged with technological leaps in communication with the telegraph and telephone, such new modes of transport as the railroad and steamship, and the consolidation of global markets—does not begin to compare with today's radical experiment. Since the 1960s, inherent tendencies in capitalism to expand markets—abetted by computers, instant communication, and falling transportation costs—and actions by governments to encourage and enable the global flow of goods and capital have unleashed a juggernaut of rapid, intense globalization, which has been drawing us all ever more tightly into a single global economy and a single human population.

The past few decades have been "days of miracle and wonder," in the words of songwriter Paul Simon, with worldwide direct-dial long-distance calls and satellites filling the air with "staccato signals of constant information." During the Earth Summit in Rio de Janeiro in 1992, I sat in the plenary session listening to world leaders promise to stem the growing crisis, while a close friend watched the same event on a television powered by a generator in a village in the rainforest on the island of Sulawesi. People in a remote corner of Indonesia could see and hear U.S. President George H. W. Bush declare to the world, "The American way of life is not up for negotiation." The United States, he made it clear, would not even tolerate questions about the status quo, much less consider altering it in some way to address the challenges of the planetary era. If television in the jungle is one sign of these globalized times, the other is certainly the phenomenon, which has developed since the late 1960s, of overnight air-express deliveries almost anywhere in the world. This degree of interconnectedness is unprecedented in human history.

Today, with the help of computers, satellites, and a speedy global transport network, manufacturers have dispersed the process of producing goods around the world, creating a global organization of resource supply and production that requires exceptional stability. This system, which values economic efficiency over security, has done away with maintaining "just-in-case" backup reserves of essentials, relying now on "just-in-time" supply policies in which raw materials and necessary components arrive only as needed and not before. This strategy, which was developed in Japan, has since been adopted around the world. Maintaining just-in-case reserves is a practice as old as human civilization and, indeed, as already noted, one of the initial advantages of moving to agriculture and permanent settlements. Now, however, warehouses with sizeable stores of parts or materials have become a thing of the past as businesses of all sorts have shifted to manufacturing in response to moment-to-moment demands and just-in-time flow of the supplies they depend on. The Dell computer company and Walmart have been in the vanguard of this trend to eliminate inventory as much as possible.

The drive to cut costs in this way has spread to other parts of the economy as well, including food. In the world of grocery retailing, the just-in-time or "lean" supply chain has been dubbed "efficient consumer response," or ECR. In this system, computers track the data from the scanner at the supermarket checkout counter. This prompts a reorder of anything that goes out the door and the producer delivers the new shipment directly to the store, eliminating warehouses and intermediate suppliers. As the food system moves to just-in-time inventory management, cities have become highly vulnerable to any disruption. Urban areas now have only a three- to four-day supply of perishable foods, and the stock of dry grocery products has been reduced by more than 40 percent.

How long would it take to run out of food if something—a global pandemic or a break in oil supplies—were to disrupt shipping? No time at all. Former British prime minister Tony Blair came face-to-face with this frightening reality in September 2000, when rising fuel prices sparked a blockade by angry farmers, truckers, and cabbies at a Shell oil refinery in Ellesmere Port in Cheshire, a protest that quickly spread to involve at least six of the nine refineries and four oil distribution depots across the country. Within two days, gas stations relying on just-in-time delivery began running out, setting off fuel hoarding and mile-long lines at gas stations. Within four days, a third of all gas stations had shut down, and some emergency services like ambulances were in jeopardy because of dwindling fuel supplies. By the seventh day of the disruption, panic buying was emptying supermarket shelves, and some stores began to ration bread and milk. The chairman of the Sainsbury supermarket chain, Sir Peter Davies, warned the prime minister that food would run out "in days rather than weeks." Because the modern economic system had abandoned the time-tested cultural practice of maintaining healthy reserves, a First World society faced paralysis in one short, breathtaking week.

From an evolutionary perspective, this kind of globalization is a dangerous move, especially in a time of growing instability and uncertainty. In merging humanity into a single population, globalization reverses a strategy that has helped ensure human survival in the long run. We have long been a global species, but one dispersed around the world in many largely self-sufficient populations with various skills and resources. This dispersal of humans in separate, culturally diverse groups around the world—the earliest version of "globalization"— helped increase the odds that some would make it through episodes of widespread catastrophe. Through most of our history, the human species has sailed into the storm in many boats. Today, through globalization, we are all becoming passengers on one *Titanic*—modern, efficient, and vulnerable to the hazards ahead.

Guided by the economic logic of efficiency and profit, we magnify our risk through such integration, even as cultural diversity erodes and we lose local strategies for survival that have withstood the test of millennia. Jamaica, which used to rely on its own dairy farmers, now imports cheaper milk from the United States. A few centers of industrial agriculture now feed a good part of the world, so many farmers in India no longer grow crops to feed their own families, but only cotton for export. Day by day, the world is losing hard-won traditional knowledge of how to farm without fossil fuels, as well as the unique and valuable traits carried in the diversity of traditional crop varieties and farm-animal breeds. Small farmers in Poland, for example, who farm without pesticides, gasoline, or genetically modified seeds, have been finding it harder to survive since their country joined the European Union in 2006. The EU's sanitary laws and mandates to encourage competition and efficiency are threatening farming methods that yield healthier food, produce little carbon dioxide, and have relatively slight impact on the surrounding countryside, which is why Poland boasts forty thousand pairs of nesting storks. Official policies ought to be doing their utmost to promote these small farms and the cultural knowledge they preserve. They are not just the past, but perhaps a bridge to the future.

Ecologists have a term for excessive integration, which makes the whole system vulnerable to any disruption from within or from the external environment: They call it *hypercoherence*. A system in this highly connected state, observes C. S. Holling, an ecologist and leading theorist of systems behavior, is an "accident waiting to happen," because "overconnectedness" increases rigidity and thus structural vulnerability. Holling's study began with natural systems, but his observations about forests led him on to a more general study of all kinds of complex systems, including human social systems and the larger dynamic of linked human and natural systems. Over time, his seminal insights into how such systems evolve and adapt grew into a theory of *panarchy*, which outlines principles of resilience and cycles of adaptive change.

The logic of long-term survival is not an efficient one. The experience of life on Earth holds lessons that directly contradict the prescriptions of such economic theorists as David Ricardo, a nineteenth-century English financier whose ideas even today underlie the faith of economists in the virtues of free trade. To improve efficiency and achieve the greatest total production in the world, Ricardo advocated the strategy of "comparative advantage," in which regions concentrate on producing the one thing or few things for which their comparative costs are lower and then acquire other necessities through extensive trade. But life has endured for 3.5 billion years on a turbulent Earth by hedging bets and seeking safety rather than by pursuing maximum production and boundless integration.

A system's structure is a key factor in whether it is rigid and vulnerable or resilient and adaptable. Whatever its other costs and benefits, today's high-velocity globalization—which aims to create a single, efficient global market and remove barriers that prevent the free flow of capital across national borders—is restructuring the human system in ways that make it more precarious. As the *number and strength of connections* increases, the system becomes vulnerable to any number of disruptions, which can spread rapidly across markets and societies. (Nothing has illustrated this danger more vividly than the way a U.S. crisis stemming from irresponsible mortgage lending practices spread rapidly to other economies around the world in 2008, precipitating a global financial meltdown.) With *speed of communication*, markets react instantly, making the global economy more volatile. Globalization is also problematic because it drives out *diversity* of all sorts—biological, institutional, technological, ethnic, and cultural, as well as diversity in language, tastes, and values—which is essential for resilience and the capacity to adapt to future changes. This homogenization is a direct result of the drive for maximum efficiency at the expense of other values. Global trade in industrial goods creates economies of scale and lowers prices to levels that smaller producers and regional markets cannot match. But taking issue with this loss of diversity entails more than romantic nostalgia. Diversity preserves options and also fosters the novelty and experimentation that are the grist for both cultural and biological evolution.

At the same time, some of the benefits of globalization allow us to better combat its consequences and aid those who have been harmed by its impact. Today's computers and communication technologies offer unprecedented opportunities for global collaboration among people eager to address the growing dangers through the spread of useful ideas and technology. It is also now possible, thanks to advances in communication and transportation, to mount relief efforts for victims of droughts, floods, and other natural disasters on a scale never before possible. But, however helpful, the positive side of globalization has done little so far to stem the opposing trends that are propelling us into ever greater danger.

The tightly integrated global economy, which makes profit and efficiency its overriding goals, increases vulnerability in all kinds of human systems. Take for example the growing dangers to the manufacturing system, to the global food supply, and to public health because of pandemic disease. A quick survey leaves no doubt that the modern human enterprise has recently gotten itself far out on fragile limbs. Efficiency begins to seem a form of rational madness.

Even routine disasters threaten the integrity of the global economy, because of extreme and risky globalization, in the view of Barry Lynn, an author and journalist specializing in globalization. "Our corporations have built the most efficient system of production the world has ever seen, perfectly calibrated to a world where nothing bad ever happens. But this is not the world we live in." As a consequence, "a sixty-second earthquake in Taiwan can nearly shatter the American economy," because the electronics industry has relied on specialized semiconductor chips made by two companies who manufactured them in the same industrial park. In a similar way, he warns, America could lose its ability to build cars and airplanes because of an epidemic in China, or see half of its supply of flu vaccine disappear because a factory closes in England. We are making ourselves vulnerable "in the name of efficiency," and he warns that the next industrial crash is only a matter of time. The sobering thing is that Lynn's analysis excludes the danger posed by a future climate shock, focusing only on the vulnerability of globalized manufacturing under normal climatic conditions today. Government and business leaders continue to pursue policies that have made human systems "more efficient than safe."

The World Economic Forum, an independent international organization of the world's business and political elites that holds annual meetings in Davos, Switzerland, acknowledges the dangers arising from growing interdependence and global supply chains. Its report *Global Risks 2007* stated that supply-chain disruptions can be "potentially catastrophic" for a company's profitability, but did not address the larger issue of how the drive for economic efficiency that created those fragile supply chains is now undermining the overall security of societies. The next year, however, the *Global Risks 2008* report finally noted the "systemic vulnerability to a supply chain failure," acknowledging that "even a relatively small supply chain disruption . . . may ultimately have consequences across the global economic system."

This tight global interconnection in a time of climate change may have far-reaching consequences for public health, as well. Rising temperatures and erratic and extreme weather promote the conditions in which infectious diseases thrive. With the growth of international travel and commerce, an outbreak in one area can become a problem halfway around the world in a matter of days. In coming decades, warns Paul Epstein of Harvard Medical School, more frequent outbreaks, epidemics, and quarantines (such as those imposed in Canada and China during the 2003 SARS crisis and in Mexico and elsewhere during the 2009 H1N1 flu outbreak) will add to the burden of societies already stressed by the disruptions of global change.

There are other dangers as well, says Dr. Michael Osterholm, director of the University of Minnesota's Center for Infectious Disease Research and Policy, who has been a medical Paul Revere warning about the need to prepare for bird flu and other possible pandemics. When a major crisis does hit, it will quickly overwhelm the healthcare system, he warns, because, like the rest of the economy, hospitals operate with lean inventories and depend on just-in-time delivery of essential supplies. In testimony before Congress, Osterholm deplored the folly of operating without stockpiles of surgical gloves, masks, IV bags, and mechanical ventilators. If countries close their borders to stop the spread of the virus, the global economic system, which depends on the rapid delivery of replacement parts for equipment, of food, and of raw materials, will abruptly stall. This failure will lead, among other things, to a shortage of antiviral treatments and vaccines, because U.S. drug companies import 80 percent of the raw materials used to manufacture their products. "Today, we have no surge capacity for any consumer product or medical service that might be needed during the 12 to 36 months of a pandemic," Osterholm wrote in a commentary in the journal *Nature*. "The pandemic-related collapse of worldwide trade and its ripple effect throughout industrialized and developing countries would represent the first real test of the resiliency of the modern global delivery system." Is there any reason to believe it will pass this test?

The globalized food production system is also at growing risk. Molly D. Anderson, a leading food systems analyst, describes it as "several disasters waiting to happen" and judges that we are unprepared even for immediately evident challenges. These include the end of the cheap oil that makes industrial agriculture and international food distribution possible and affordable, as well as the rapid loss of genetic diversity in food plants and farm animals just at the time we will need it most. In the last century, the world has lost three quarters of the genetic diversity in crops. At the moment, we are spending far more to protect wild animals than to protect threatened breeds of livestock that contain a diverse gene pool, which is vital to future food security. The world is losing a breed every month, and roughly 20 percent of the more than 7,600 breeds in the United Nations Food and Agriculture Organization global database of farm animal genetic resources is at risk of extinction.

Globalization of livestock markets is "the biggest single factor" driving this loss, according to the FAO. Modern agriculture has placed a premium on breeds that produce high volumes of meat, milk, and eggs, while traditional farmers have valued animals for more than high output. The Kuri longhorn cattle of Africa's Sahel, for example, are equipped to deal with weather extremes: They can cope with heat and drought and are strong swimmers in a flood. Such tough breeds have withstood the test of time, disease, and erratic weather. But today they are dying out as farmers replace them with specialized, high-yield breeds, such as black and white Holstein-Friesian cows, which are now found in all regions of the world. In Uganda, this switch proved devastating to farmers in a recent drought. They lost entire herds while farmers who had stayed with the indigenous Ankole cattle made it through the dry period because this hardy breed was able to walk long distances to the remaining watering holes.

Scientists are also raising the alarm about the dangerous narrowing of the gene pool *within* these few super-breeds, such as Holstein-Friesians, that are coming to dominate the human food supply. In the quest for maximum production, breeders have taken to artificial insemination using the sperm of a few prominent stud bulls, which, as their semen is shipped internationally, are fathering cows all over the world. Viewed through the lens of population genetics and the measure of "effective population size," all of the Holstein-Friesian cows in Germany are so inbred that their genetic diversity is equivalent to only fifty-six individuals, while the Japanese black cattle have an effective population of seventeen.

The diversity of plant varieties and genetic resources is also declining rapidly as commercial agriculture has replaced small farmers and people have moved to the cities. In Poland, traditional farmers grow thousands of seed varieties that are not found in commercial agriculture. Elsewhere in Europe, backyard gardeners have saved seeds and passed down heirloom varieties of vegetables and fruit over generations, but these are being lost because their city-dwelling children have neither the interest nor the time to preserve this legacy. Scientists in Italy and else-where are in a race to collect as many of these unique varieties before their seed-saving guardians die and the plants—which differ signifi-cantly from the varieties favored by large industrial growers—vanish with them. This is more than a loss of plant varieties; it is an alarming loss of options as plant breeders try to develop crops that are tolerant to drought, heat, and rain as climate change alters growing conditions.

Growing instability in planetary systems combined with increasing vulnerability in human systems—these are the double dangers that threaten our complex, interconnected global civilization. The plane-tary emergency is not simply a problem of climate change and the need for an energy system free of fossil fuels. The current trajectory, accelerated by high-velocity globalization, is altering the Earth's over-all metabolism even as it reduces human resilience and our capacity to adapt to future changes.

How will possible climate surprises ripple through a volatile global economy? How susceptible are we to the danger of a cascading collapse of social and economic systems? The possible answers are difficult even to admit. Because of excessive global integration, the shock of a re-gional climate shift—for instance, the sudden onset of catastrophic drought that would make California and parts of the Midwest grain belt uninhabitable—could precipitate a crisis that would ripple through the political system and the global economy and perhaps plunge the world into a global depression. The danger of such a precipitating event is all too real, since the past 150 years in California have been by far the wettest of the millennium. In the longer term, extreme and per-sistent droughts have been the norm. In the Medieval Warm Period from roughly A.D. 800 to 1300, rainfall in California fell to only 25 per-cent of what has been the recent norm and stayed at that level for more than two centuries—a grim period when 50 percent of the West stag-gered under a megadrought. During this time, sand dunes blanketed vast areas of the Great Plains stretching all the way from Texas and Oklahoma through the heartland to Saskatchewan and Manitoba. Ed Cook, a scientist who has worked on reconstructing these droughts through tree ring records, has described the prospect as scary, adding, "You take water away, and the semiarid West just depopulates." In such a scenario, *the gravity of the economic damage could far exceed the scale of the environmental insult* and leave public officials without the resources to deliver emergency assistance; to provide water; to manage refugees; and to fund the permanent relocation of large populations to habitable regions. As things now stand, one of the first surprises of climate change might be how quickly global industrial civilization unravels in the face of relatively minor disruption.

With the possibility of abrupt change ahead, the pursuit of "sustainability" or "adaptation" to a warmer, stormier, or drier world won't provide our children and grandchildren with the necessary defenses. In order to improve the odds for those who will be making the dangerous passage through a time of turmoil, we must *expect surprise* and think about "survivability"—the task of shockproofing our human systems as much as possible against the whole range of possibilities that we face. Such preparation is without question as urgent a priority as reducing the emissions of the gases that are driving climate change. If we do not do what we can to insulate human systems, a period of wild climate change could leave us without the social and economic resources to maintain basic services and institutions necessary for organized human life. These range from maintaining food and transportation systems and safeguarding water supplies to burying the dead. Resilience has to trump efficiency, and we must foster *adaptability*—the social capacity to respond to future changes—above all.

Adaptation, which can be preventive or reactive, is a separate matter, involving specific measures to address new physical challenges like rising sea levels, more intense rains and flooding, and more prevalent drought. Those facing threats can respond to vulnerability beforehand

and therefore suffer far less damage than those who react only after a disaster in order to protect themselves from a recurrence. The Dutch provide a good example of preventive adaptation, for they are already at work on ways to adapt to heavier winter rains. They are lowering dikes, which now hold back water along rivers and estuaries, in several dozen spots to create flood zones and constructing amphibious houses that can float upward as the rivers flood and waters rise. We might also adapt by developing drought-resistant crops, building higher bridges to rise above higher flood surges, and as sea levels rise, by beginning a planned retreat from low-lying coastal areas to higher ground. The United States, however, has done little to anticipate and therefore limit damage from rising seas, more intense hurricanes, or other foreseeable consequences of climate change. According to a recent survey, "Those organizations in the public and private sectors that are most at risk, that are making long-term investments and commitments and have the planning, forecasting, and institutional capacity to adapt, have not yet done so." But, as its author, Robert Repetto, acknowledges, there are significant obstacles to adaptation, notably uncertainty regarding future climate change at regional and local scales, about the frequency of extreme weather events, and about the ecological and economic impacts. The questionable assumption that the United States can and will readily adapt to climate change has had political consequences, for it has supported the view that the United States has much to lose and little to gain from joining the Kyoto Protocol.

But how would we begin to shockproof the human system? Preventive adaptation, to the extent that it is possible to anticipate threats correctly, may help communities weather extreme events and continue to function during disasters and disruption. But protecting the society as a whole is another matter. The ecologist Simon Levin of Princeton University has analyzed the design features of natural systems that have enabled life on Earth to endure through repeated catastrophes—evolutionary lessons that are relevant to human society as well. The dynamics of human socioeconomic systems share strong similarities with ecosystems and the overarching Earth system; they are all "complex adaptive systems." So what are the secrets of survival on a volatile Earth? Chief among them, according to Levin, are redundancy, diversity, and a modular structure—strategies that stand in stark opposition to the thrust of efficiency-driven globalization.

The Earth as a whole, the great metabolism, has survived because diverse species play the same or similar roles in ecosystems, providing what Levin describes as *functional redundancy*. This kind of diversity goes beyond the "biodiversity" measured in the number and variety of species. The presence of very different species that do the same job is insurance; it provides alternative ways to keep the system going in the face of change. If, for example, a species that is now a major player falters as the climate becomes drier, another species may be more tolerant of the new conditions and able to carry on such critical tasks such as nitrogen fixation or carbon recycling. Most of the species playing particular ecological roles may not be the star performers; they may be less efficient or successful than those that dominate for now, but they could become the principal players when the conditions for life shift.

Take, for example, the changing fortunes of *Emiliania huxleyi*, the tiny armored coccolithophore plankton described in chapter 2. This tiny organism, which is individually too small to be seen with the naked eye, is one of planetary importance. Today, this species, which its scientific fans call *Ehux* for short, is the most abundant species of coccolithophore on the planet and famous for its massive blooms that explode across the deep blue face of the Earth's oceans in swirls of dazzling turquoise and white. *Ehux* helps maintain the planetary metabolism in several ways. As mentioned earlier, it plays a role in the global cycling of sulfur by emitting the gas dimethyl sulfide. It also helps maintain the heat balance of the planet: Dimethyl sulfide creates particles that foster the formation of clouds that shade the oceans, and

ast, white *Ehux* blooms at the water's surface reflect heat and
t back to space. *Ehux* is also key in the process that transfers carbon dioxide from the atmosphere to the ocean and then locks it away
in deep-sea sediments. It does this by incorporating the carbon present in seawater into its shells; because the shells are heavier than the
surrounding water, they help carry the carbon to the deep sea. (Ocean
fertilization, discussed in the previous chapter, encourages plankton
blooms, but the species that proliferate do not necessarily deliver carbon as reliably to the ocean bottom as *Ehux* does.) The geological
record indicates that *Ehux* is a newcomer to this starring role in the carbon cycle, becoming preeminent among the coccolithophore species
only since the end of the last ice age. Before the last glacial period, another coccolithophore genus, *Gephyrocapsa*, dominated the scene,
forming the big blooms on the world's oceans and transporting carbon to the depths. Scientists do not understand exactly what environmental changes sent *Gephyrocapsa* into eclipse and brought *Ehux* to
the fore. *Gephyrocapsa* did not disappear; it is still around, but it no
longer plays a significant role in carbon transfer. Thanks to functional
redundancy, however, the carbon cycle has continued despite the altered conditions and changing cast. The great metabolism has cycled
on without interruption.

Modular structure and compartmentalization have provided another type of insurance. The species in an ecosystem may be connected to one another, but they are not thoroughly interdependent or
integrated into a seamless, single system. On the contrary, an ecosystem is a web made up of clusters of species, which interact strongly
within the cluster but have limited connection to other clusters.
"When interconnectedness is reduced," Levin observes, "distur-
bances are contained." Systems with this kind of compartmentalization are insulated from the "cascading collapse" that occurs when a
disturbance in one part spreads rapidly and threatens the stability of

the entire system. This structural feature reduces "the potential for catastrophic avalanches of extinction" and at the same time provides the means of recovery when the crisis has passed.

With globalization, Levin warns, the modularity of the human system has broken down, "compromising and threatening the resiliency of humanity." We need serious reflection on the wisdom of excessive integration in a period of growing instability. We need to incorporate the secrets of survival—functional redundancy, diversity, and modular structure—into our human systems. Or rather, reincorporate them. Until recently, the human population—scattered across the globe in distinct cultures—had structural features similar to those shaped by evolution in the Earth system as a whole.

The current trends in our global civilization hold dangers even in ordinary times. In these extraordinary times, they could well put the survival of organized human life in jeopardy within the lifetime of a child born today. The responses we've mustered so far to this planetary emergency have been reactive and tactical, in that these efforts, which have been far too feeble, have focused on symptoms and attempted to remedy them with new treaties, laws, programs, incentives, or technologies. If our societies are to stand any chance of making it through the twenty-first century intact, it is imperative that we step back and think first about strategy, including the crucial question of objectives. Our primary aim, in my view, must be to preserve through this century of deep change the possibility for coherent, civilized human existence. Yes, we likely will be in that much trouble by 2100 if we don't depart sharply and quickly from the current trajectory. The means for doing this is the two-pronged strategy I outlined earlier: a rapid retreat from humanity's great global experiment, which is unhinging Earth's metabolism, and a concerted pursuit of survivability by increasing the resilience of our communities and institutions.

Survivability is nothing new or radical. It is, in fact, the oldest life logic and the way that Earth and our human ancestors have managed to endure. Today, the tried-and-true practice of striving for sufficient safety rather than maximum production survives primarily among poor, small-scale farmers around the world, who know firsthand how critical hedging bets is to survival. If the goals of security and redundancy sound extreme in this globalized age, then it testifies to the radical nature of the modern era's values. How mad is it that the current order is now dispensing with the very rationale for civilization: the desire to enhance security through warehouses and backup supplies that could carry people through lean times? We have survived in the long run by keeping our options open and hedging our bets, not by running our societies as if we lived in a world where nothing bad ever happens.

Now finally we come to the question of tactics. I don't aim here to provide a tactical manual for the coming century. Indeed, the failure to rise to this crisis does not stem from any lack of creative ideas or expertise about how to begin realizing them. The failure has rather been one of courage, of political, intellectual, and moral leadership, clarity of objective, long-range vision, and commitment to tomorrow and the day after. To have a fighting chance, we must be absolutely clear about where we need to be headed and make choices that further the two-pronged strategy of reducing the human burden on planetary systems and reducing human vulnerability to this disruption. It won't be enough to "green" the status quo—for example, by switching to electric cars—to find new ways to profit through "green businesses," and to create "green jobs," if economic growth outstrips gains through efficiency and the scale of the human enterprise continues to expand.

There is great importance in how we frame the challenge. The task may seem overwhelming—indeed, impossible—if we ask it in the abstract: How do we redirect an entire civilization? It becomes less intimidating if we ask, What are the things making us especially vulnerable? And how do we address those vulnerabilities and make ourselves safer in the face of the unexpected? This latter question has a short answer: Reverse this excessive integration, diversify, and disperse the production of food and essential goods, so that we no longer keep all our eggs in a single global basket. The longer, more concrete answer is a matter of tactics. Many are already thinking about how to improve safety in areas such as agriculture and manufacturing and advancing specific measures to accomplish this. Once we take our vulnerability seriously, there will be no lack of good, realistic ideas about how to improve human security in this new historical landscape.

The prospects for some retreat from the excesses of globalization are, moreover, improving for a variety of reasons, including the current global financial crisis, the likelihood that the days of cheap oil are over, and growing political and environmental concerns about lost jobs, climate change, and food safety and security. The collapse of world trade talks in Geneva in August 2008 was clearly a sign that the globalization juggernaut is no longer as inevitable as many claimed in the past.

In Barry Lynn's critique, the current manufacturing system is fragile in part because of too tight an interdependence among nations. He therefore offers a number of far-from-radical recommendations that would begin to make it less subject to collapse—remedies that embody the same fundamental principles of redundancy and compartmentalization outlined by Simon Levin. The government could, for example, enhance safety by mandating diverse sources of the necessary raw materials and services used in manufacturing. He also recommends requiring firms to rely, at all times, on two or even three sources for all components and services from suppliers in two or more nations. If government regulators required companies to disclose their sourcing and supply-chain relationships, it would give investors the information they need to avoid firms that take unnecessary risks. Some business gurus, such as Jeffrey E. Garten, the former dean of the Yale

School of Management, judge that the growing risk of the just-in-time system will encourage companies to return to policies of redundancy, such as warehouses, multiple sources of supply, and perhaps multiple production sites.

Lynn challenges head-on the idea that the forces creating the current global economy are inevitable and unstoppable, that this development is out of human control. "There is nothing ineluctable or necessary about today's global industrial system, " he writes. This kind of "deterministic imagery serves only in the writing of tragedy. A deterministic economics serves only to ensure that someone other than us shapes our lives." In fact, the current economic interdependence and international trade policy have come about through deliberate government action in the postwar era and most recently in the policies of the Clinton and Bush administrations. Our current predicament is the result of a choice. In the light of new dangers, we can now make different choices if we see fit.

Over the past two decades, there has been an explosion of interest in agriculture and food issues and growing activism on many fronts in the United States and internationally. This new food movement is rich and many-faceted and, as with Lynn's proposals for manufacturing, runs counter to excessive integration, generally aiming to promote a food-supply system that is primarily local or regional. These efforts include the Italian-born Slow Food movement to preserve the cultural legacy of food and cuisine; the growth of support for organic, local, sustainable farming, including farmers' markets; and the transnational "food sovereignty" campaigns that are battling policies—promoted by the United States, the European Union, the World Trade Organization, the World Bank, and the International Monetary Fund—aimed at globalizing food production. Food sovereignty advocates, who have rallied around an objective first formulated by an international peasant farmers movement, La Via Campesina, emphasize national control over food policy, and farming directed toward self-sufficiency rather than the production of export crops. Activists have been engaging in these issues on all levels, from the local and regional to the national and international. In the United States, a broad coalition of activists made some headway in the 2008 Farm Bill, winning modest support for programs that encourage small to mid-size farmers and organic agriculture in legislation otherwise focused on the interests of corporate agribusiness.

Food sovereignty advocates have focused their energies recently on a campaign against the effort of the WTO to bring agriculture fully into the free-market trading system. In July 2008, farm and food activists celebrated when WTO talks collapsed, in part due to concerns of developing countries that their local agriculture would suffer from a glut of cheap, industrially produced food. Proponents of international food trade have opposed food sovereignty, contending that it hampers the free flow of trade that lowers prices and makes food more accessible to all. Because some countries cannot grow enough food to feed their populations, some trade in food is inevitable. The question is rather what kind of trade, who controls it and benefits from it, and how that trade fits into the broader picture of local and regional agriculture.

So far, concern about overall human vulnerability in the face of global change has played a minor role in the burgeoning food movement, but it adds new urgency to these efforts. In October 2008, the International Commission on the Future of Food and Agriculture brought the issue of vulnerability to the fore with its "Manifesto on Climate Change and the Future of Food Security." This document critiques how industrialized agriculture fuels climate change and is vulnerable to its impacts. It makes the case for local, sustainable food systems.

Many individuals, nongovernmental groups, and governments around the world are already working to minimize risk through diversification of the crops, cultivars, and the places where food is grown. The

Global Crop Diversity Trust, an independent international organization, is now raising a $260 million endowment to further this work, accepting donations from everyone from the Bill and Melinda Gates Foundation to small donors. In September 2008, the trust launched a project that will search national seed banks for "climate-proof" varieties that are tolerant of extreme weather—floods, droughts, and temperature swings. Governments, foundations, and individual volunteers and contributors should bolster these efforts and others, which are often scattered, small-scale, and inadequately funded, in order to safeguard prime farmland from development; to preserve heirloom varieties and wild ancestors of food plants and endangered breeds of farm animals; and to pass on cultural knowledge of nonindustrial farming, traditional skills of preserving food without refrigeration, and preindustrial methods for producing basic tools and goods like buckets, barrels, and plows.

Maintaining as many options as possible is imperative for survival, yet governments give it low priority. In cases where they do provide some funding for such efforts, they typically also promote trade or agricultural policies that drive out diversity in the name of efficiency. The impact of EU policies on traditional Polish farmers illustrates the shortsightedness and perversity of such policies.

One outstanding example of foresight and insurance for future generations is the "doomsday vault," a $3 million seed storage center that the Norwegian government is building as a service to the world community in the remote Svalbard archipelago above the Arctic Circle. Though other seed banks exist around the world, they often lack secure funding and could be lost in natural disasters, wars, and other catastrophes. The Svalbard International Seed Vault is a backup, which will store 3 million seed types in a chamber surrounded by permafrost and rock, ensuring that the seeds will remain frozen even if the electricity fails. The Norwegian government will pay an additional $150,000 a year to maintain the vault, and the Global Crop Diversity

Trust will support its operation and management—a trivial investment, all in all, for providing long-term security for a civilization that depends on agriculture for its survival.

In the coming decades, we need to confront the question of how to feed the world without cheap fossil fuels, since industrial agriculture cannot function without oil and gas. Here again, grassroots groups are already sprouting, many of them part of the international Relocalization Network, made up of 150 local groups around the world that seek to think, educate, experiment, and agitate about making the transition to a postcarbon world. Independent writers and thinkers like Richard Heinberg, who is now a fellow at the Post Carbon Institute, have been in the forefront, asking important questions that still get too little serious attention within the mainstream media about the vulnerability of the food system to fuel shortages and skyrocketing prices. He has become well known for his provocative and well-reasoned essay "Fifty Million Farmers," which lays out the reasons why the deindustrialization of agriculture may be unavoidable when the flamboyant era draws to a close, hence the need for many more farmers. In working to reduce demand for fossil fuels, to develop local and regional food production, and to foster social solidarity through events like solar panel raisings in the spirit of traditional barn raisings, local groups reduce stress on planetary systems and enhance survivability. Activists are also exerting influence on local governments by proposing community-wide measures to reduce the demand for fossil fuels—ideas that have increasing traction with rising fuel prices. These include better public transport, projects to make bicycle travel safer and easier, and investment in alternative and diverse sources of energy, such as wind, solar, and geothermal.

On other critical fronts, however, there has been no progress at all and probably won't be without government leadership. Chief among these is readiness for emergencies in a world where bad things do happen and where future disruptions may be *systemwide* rather than regionally limited events like a hurricane on the Gulf Coast, a crippling ice storm in the Northeast, or a terrorist attack on a major city. The U.S. Department of Homeland Security (DHS) promotes a National Preparedness Month, and other federal agencies and some state and local governments do have emergency plans, but the planning clearly does not go beyond limited, short-term emergencies. The scope of anticipated disasters is evident in a 2008 DHS news release: "From wildfires and earthquakes in California, to hurricanes and tropical storms along the Gulf Coast, to flooding in the Midwest, recent events remind us more than ever that we must prepare ourselves and our families for a disaster. This is the time, each year, when every American should ask the question, 'Am I ready?'"

If the federal government will not begin to address the question of how to manage some significant, system-wide disruption, then the association of governors and mayors should take up this critical matter of social defense. After all, they will find themselves on the front line if things fall apart. Given the absence of surge capacity in essential systems and the lack of backup stores of food and basic medical supplies, it won't take much to send communities and countries into panic and chaos. Whether by direct government mandate or by various forms of incentives, it is in everyone's interest that supermarkets, medical supply companies, pharmaceutical companies, hospitals, and other essential institutions adopt just-in-case practices, with warehouses and backup supplies sufficient to meet basic needs for some specified period of time. It seems absurd that the Pentagon devotes a great deal of money and energy planning for long-shot military threats, but the U.S. leaders can't muster the attention and resources to prepare for events—bird flu quarantine, a megadrought in the western half of the United States far worse than the Dust Bowl, a cut in Middle Eastern oil deliveries—that are altogether possible. At the moment, it would appear that our public officials' only response for a six-month or yearlong emergency would be martial law. Without well-developed plans ready to provide basic necessities and prevent social breakdown, that would seem the only choice.

Hurricane Katrina demonstrated the precariousness of the world oil supply when it not only shut down oil production in the Gulf of Mexico but also disabled 30 percent of U.S. refinery capacity. The United States does maintain a large stockpile in its Strategic Petroleum Reserve, but here was the hitch: The backup supply was all crude oil; none in the refined form—gasoline, diesel, jet fuel—that actually runs the world. Moreover, even before the storm, refineries were barely able to keep up with the demand for oil products; the world system has no excess capacity. It proved fortunate that European countries hold some of their emergency reserves as refined products, so the International Energy Agency was able to direct twenty-five tankers carrying gasoline to the United States and avert turmoil in the fuel market. Given the current dependence on oil, federal energy officials need to address the risky concentration of refineries in a hurricane corridor, the absence of surge capacity in oil refining worldwide, and the form in which we keep oil reserves.

In the longer term, governments need to provide incentives for a transition to diverse sources of energy, especially those forms that can be produced on the local or regional scale without relying on resources imported from distant places. The goal in this transition must go beyond finding replacements for coal, oil, and gas and meeting increasing energy demands. This is an opportunity to redesign the structure of the energy system based on Levin's principles for greater security.

The UN Food and Agricultural Organization has warned that the world is facing a global food crisis driven by volatile oil prices, growing demand from countries like India and China, increasing excessive speculation by institutional investors in commodity markets, the shift by U.S. farmers from cereals to biofuel crops, and extreme weather. In late 2007, global food reserves dipped to the lowest level in twenty-five years, and the following summer, the prices for staple foods reached record highs. The global cereal stock, which has been dropping for more than a decade, now stands at around fifty-seven days, an alarmingly thin defense against hunger and famine. The world is already seeing protests and food riots because of soaring food prices and thus a growing danger of political instability.

As a general rule, it is short, sudden price increases that do the worst social and economic damage. Based on his study of such episodes in the past, Brandeis University historian David Hackett Fischer recommends that countries establish commodity reserves for basic resources and agricultural products on the model of the petroleum reserve. He notes that such reserves and a stand-by system of price controls, which have worked well in the past, provide basic tools for preventing destabilizing price surges.

It is astonishing to learn that the policies of the United States and international bodies like the World Bank and the IMF have discouraged the practice of maintaining grain stockpiles that could provide insurance and dampen the volatility of food prices. Under the 1996 Farm Bill, the federal government dismantled publicly sponsored grain reserves, based on the view held by some agricultural specialists that large government stockpiles are both unnecessary and counterproductive and entail storage costs. The assumption, according to Dr. Daryll Ray, who directs the Agriculture Policy Center at the University of Tennessee, was that "the commercial sector would dependably provide whatever stocks we would need, but that is not the case." Ray noted that federal corn stocks provided the necessary cushion and kept prices stable during years of reduced harvests in 1983, 1988, and 1993. In the spring of 2008, a coalition of farm and religious groups sent an open letter to Congress, which warned that the lack of reserves increases the risk of higher prices and hunger and called for the creation of a strategic grain reserve, arguing that the lack of domestic and international grain reserves has fueled the global food crisis. The nonprofit consumer organization Food and Water Watch blames

World Bank policies for pushing many African countries to abandon their reserve programs—a move that has made them dependent on imported grain, which doubled in price between 2007 and 2008. Kenya and Malawi, for example, abandoned their reserves altogether at the bank's urging.

In July 2008, the World Bank revised its official stance on this issue, as its president, Robert Zoellick, urged leaders of the G8 to back a food-reserve plan developed by the International Food Policy Research Institute. The response fell far short of what Zoellick proposed. In its statement at the end of the summit, the Group of Eight economic powers did little more than call on those with sufficient stocks to release some of their reserves. This is another appalling failure of leadership.

The time is ripe to question the economic dogma that has set the course for national policy and shaped the larger world through the policies of the WTO, the World Bank, and the IMF. At the heart of the clash between efficiency and prudence is a deeper question about the relationship of the economy to the larger society. During the modern era in the West, as the economic historian Karl Polanyi observed, "human society became an accessory of the economic system." Whatever the merits of this arrangement in the past, and there have been many, its dangers at this historical juncture are becoming all too apparent. Characterizing climate change as "the greatest and widest ranging market failure ever seen," Sir Nicholas Stern, who headed the Stern report on climate change and the economy for the British government, made it clear that the human future cannot be left to markets. Moreover, the perils of excessive integration became all too apparent again in late 2008 as a mortgage crisis stemming from irresponsible, often fraudulent lending practices in the United States quickly mushroomed into a global financial crisis. The cause wasn't simply "greed on Wall Street" or deregulation policies that were part of the reigning market fundamentalism. It became a global crisis because of "tight chains of financial interdependence," which make the system vulnerable to the kind of cascading collapse described by Simon Levin.

Under the weight of recent events, the fundamentalist faith in markets and globalization has been deeply shaken. While the Stern report made the case for taking action now on climate change because inaction threatens the economic future, it did not address the larger question its findings raised. If markets *fail* in the face of life-and-death challenges like climate change, how much can we rely on them to organize human efforts, and how can we prevent further catastrophic failures? The climate-change debacle and the near meltdown of the financial system raise questions about the *structure* of the human system. This is a profound problem that can not be left to economists or any group of specialists. We need to develop not just new ideas about how to fix our existing economy, but new ways of thinking about how to meet the physical needs of our societies, perhaps through think tanks with a holistic bent and new integrative institutes that bring many different perspectives and kinds of expertise to problems. Perhaps in this way we can develop market arrangements to create wealth without undermining the security of the larger society. In reforming policy, the aim should be to impose requirements on the structure of the system, not simply to attempt to regulate the behavior of those working within it. A compartmentalized structure will limit the damage even if everyone in that part of the system is behaving badly and the regulators fail in their duties.

If deep change is in the offing, now is the time to be asking the *biggest* questions. Few are more fundamental than how the economic system fits within larger objectives, such as the integrity of the Earth system and human security in the planetary era. How can we make the economy an accessory to human society and harness its energies to a mission far more urgent than the accumulation of wealth? Like the

Earth system, human history has periods when it moves by leaps rather than by small, incremental steps. These rare moments arrive like earthquakes, shattering the existing order and shifting the landscape of possibility. They are typically times when values shift and human societies establish new aims. "We are in a phase when one age is succeeding another," as Vaclav Havel put it, a moment "when everything is possible." When that moment of possibility suddenly presents itself, we need to be ready to seize the opportunity to define the values for the coming era and to reorient our societies to aims appropriate to the times and circumstance.

Crises bring out contradictory impulses in the human heart. They can be times when people pull together or when the social order falls apart. In extreme times, what some have called social capital—the capacity for trust and cooperation that has played a powerful role in human success and survival—often proves far more precious than capital of the financial kind. This is still an important asset in groups living in harsh and unforgiving environments, like the Gabra in the Chalbi Desert in Kenya, who by custom freely lend camels to others in their group and even to total strangers in times of drought and dire need. The "camel bank" and the social network it fosters provide security that no amount of money can command. This point was made chillingly by a sixth-century Japanese king, Senka, whose words were recorded in the *Nihon Shoki* (Chronicle of Japan) in A.D. 536, a time when societies around the world were engulfed in climate chaos brought on by a catastrophic volcanic eruption in Indonesia. The gases, dust, and smoke created by the blast had blotted out the sun for months and shrouded the planet in a cold, terrifying gloom. Crops failed in many parts of the world. Money became worthless. "Yellow gold and ten thousand strings of cash cannot cure hunger," King Senka despaired. "What avails a thousand boxes of pearls to him who is starving of cold."

We must do our utmost to enhance social capital. This is already happening at the local level in many places where, for example, people are working together to improve the quality of school lunches, to teach children to grow food, and to rally their communities behind initiatives to encourage local markets. But there is an equally pressing need for social capital in the global arena, for much in the decades ahead will depend on how we deal with worsening scarcities.

The British historian John Gray warns that we stand on the brink of "a tragic epoch, in which anarchic market forces and shrinking natural resources drag sovereign states into ever more dangerous rivalries." There is an alternative—the pursuit of coexistence and cooperation through a regulatory framework to manage limited and declining supplies of fossil fuels, food, and other critical resources in the coming decades. The G7, which includes all the leading economic powers except Russia, could take the lead and agree, for example, to begin preemptive cuts in consumption now, which would allow for an orderly transition out of the era of cheap oil and other essential resources. This might well save us from a global depression precipitated by repeated price shocks. Taking such steps, as journalist David Strahan put it, "demands a recognition that *business as usual* means collective suicide"—something world leaders have yet to admit.

We must also confront the possibility that the chaos of the coming century may prove too much for our current global civilization. I'm by no means forecasting that civilization is headed for collapse, but given the great uncertainties in both the environmental and the human realm, a dark age is one possible scenario, and this time it would be a global phenomenon, not a regional event, as it was after the Roman Empire disintegrated. For this reason, we should also be thinking about how to convey essential cultural legacies through a time of disintegration when today's technologies and even literacy could be largely lost. This cultural kit should include vital knowledge gained in the modern era. My list would begin with the fact that all

Earthly life shares a common ancestry and the discovery that microbes invisible to the naked eye—not curses, bad air, or evil spirits—cause disease. Preserving through a time of cultural breakdown knowledge about safeguarding water, basic sanitation, the importance of hand washing, and how to make soap could help prevent a good deal of illness and suffering. But equally important will be to pass on knowledge of humanity's diverse historical and cultural experience, including the truth that there are many ways to be human besides the modern way, and that flexibility is the very essence of human nature.

The historical lesson of how ancient learning survived through times of chaos and illiteracy might guide an international project to develop recommendations. A body like the United Nations Educational, Scientific and Cultural Organization (UNESCO) might convene such an effort and consult with national and international groups in the arts, sciences, and humanities. In the United States, the American Association for the Advancement of Science and similar broad organizations could foster this discussion. Like the doomsday seed vault, this would not be an expensive undertaking, but if and when this resource is needed, it may help our descendants in the recovery.

One final point cannot be emphasized too much. "Survivability" is far different from "survivalism." Retreating to a plot of remote land with a solar panel, a supply of seeds, and a gun is a poor strategy for many reasons, including the uncertainty about how climate or other forms of global change will affect any particular region. C. S. Holling offers strategic advice about how to live in such a time of change: "Do not try to plan the details . . . the only way to approach such a period, in which uncertainty is high and one cannot predict what the future holds, is not to predict, but to experiment and act inventively and exuberantly via diverse adventures in living." Hopeful experiments are already under way around the world, because many people, who recognize that the current way of life is approaching a dead end, are seeking new ways forward. Nothing will be more precious in the decades ahead than myriad options and well-developed, clearly articulated alternatives.

Survivability must not be focused on individual survival apart from larger social and economic systems, but on redesigning and insulating social systems so they can better withstand disruption and shock and evolve. The aim is to safeguard human knowledge and the institutions that give us the *capacity* to respond to changing circumstances.

Like Noah, we must begin to prepare for this dangerous passage well before the storm.

8

A New Map for the Planetary Era

If survivability ranks at the top of immediate concerns, the urgent question in the longer term is humanity's place within the larger world. As cultural beings, we make our way through the world guided by shared ideas about ourselves, about the nature of the world we inhabit and act upon, and about our relationship to the Earth and the planetary system that sustains its surging life—an understanding that constitutes the foundation of our cultural map. Humans inevitably view nature "through a screen composed of beliefs, knowledge, and purposes," as pioneering ecological anthropologist Roy Rappaport observed, so the diversity of human cultures throughout history has created a wide variety of cultural maps, some of them quite bizarre by modern standards. But all cultures face the same critical problem: "the discrepancy between cultural images of nature and the actual organization of nature." Whether a culture's explanation of the world is correct in a scientific sense may not be important, Rappaport noted; what matters is whether its cultural rituals and beliefs guide behavior in ways that allow the group to survive in its particular circumstances.

The ominous trends and surprises of the planetary era have raised fundamental questions of human existence anew. In ordinary times, such "big questions" generally don't arise because an inherited cultural map provides satisfactory answers for the time and circumstances. One can meet the world confidently, armed with a set of basic assumptions that frame a civilization's approach to life. This time, however, is not an ordinary one. Real-world events have disproved modern civilization's fundamental assumptions about nature and human power—a "ground truth" that leaves little doubt that our cultural map is not only obsolete in this uncharted historical landscape, but dangerous. So the big questions are back, urgently demanding new answers.

What confronts us is not simply an "environmental" dilemma, an "energy" or "fossil fuel" problem, or a "climate crisis"; the changes in nature are but symptoms of a profound "human crisis" that cuts to the heart of our civilization. Scientific diagnosis and technological remedy may help allay physical symptoms for a time, but they can't cure what ails us.

Rethinking the big questions is not a philosophical luxury, but rather a practical matter that bears directly on the questions we ask as we grapple with this long emergency and debate the strategies to pursue. A new cultural map is needed—one that can help orient us in these unprecedented circumstances and provide direction and coherence to human efforts at this critical historical moment. With her characteristic earthy wit, the British philosopher Mary Midgley compares philosophy to plumbing. Like water and sewer pipes, the grounding premises that support a civilization—the philosophy coded in its cultural map—are indispensable, yet both function out of sight and out of mind until something goes wrong. Then when it becomes impossible to ignore the bad smell, there is no choice, she counsels, but to pull up the floorboards and take a hard look. We have been slow to recognize that something has gone seriously wrong with our civilization itself. We desperately need to understand how and why the modern way of life has become deeply problematic. The routine business of our civilization is threatening its own survival, and by putting Earth's living system in jeopardy, it also risks foreclosing the conditions for *any* civilized life. To find our way in the planetary era, we will need a fundamentally different view of the world—one that emphasizes the very things our current map ignores.

Seen in the broad view, this planetary emergency is a crisis of context. Modern Western civilization has for the most part approached nature as a one-way conversation in which humans expect to do all the talking. According to the cultural map that has guided us in the modern era, nature is not an actor, nor is it expected to respond in any significant way. Even those protesting the damage caused by exploitation often portray the Earth as a passive, fragile victim rather than as a formidable force that might strike back. According to the modern map, human agency is all. The solid world, this physical context that surrounds us, is irrelevant beyond its use as a stage for the human enterprise and a storehouse of raw materials for the global economy. Therefore, questions of how human actions fit in the larger world get only cursory attention, if they arise at all. Even after the dangerously close call with the ozone layer, this civilization has hurtled onward, still seemingly oblivious to potentially fatal matters of context. The cultural historian Thomas Berry characterizes this profound disregard as cultural autism. For this reason, much of the behavior considered normal by the current global civilization appears pathological in light of our growing emergency—the unquestioned pursuit of exponential economic growth, the celebration of greed, wildly excessive consumption, radical individualism, overweening belief in human power, refusal to acknowledge limits, blind faith in salvation by technology, and the primacy among values of profit and efficiency.

It would be wrong, however, to take this manifestly self-destructive approach to the world as evidence of some fundamental deficiency in human nature. Such articles of modern faith as the notion of *Homo economicus*—which asserts that humans are by nature "economically rational" and have a basic drive, as Adam Smith put it, to "truck and barter"—are distinctly at odds with the way humans have engaged with the world and with one another through most of our history. As the French anthropologist Marcel Mauss famously declared, "It is our western societies who have recently made man an 'economic animal.'"

The modern emphasis on accumulating wealth stands in sharp contrast to the widespread and long-standing cultural practice in premodern societies of organizing social and economic life around cycles of gift giving.

Assuming one's own culture embodies human nature and reflects a universal "common sense" is apparently a common mistake. Just as Western economists would justify the greed and acquisitiveness of market-governed societies, an elder of the Weyewan on the Indonesian island of Sumba explained why his culture's social and moral economy is the *natural* way to organize a society: "It's the custom of life everywhere on earth [i.e., in his society] that we exchange our belongings and we don't hang on to what we have. It's our custom to exchange," he told anthropologist David Maybury-Lewis. "We exchange favors, we exchange meat and we exchange labor . . . this after all is the only way we can reach our goals. . . . If it wasn't like this, what else could we do?" In contrast to exchange in the impersonal marketplace, this type of traditional exchange fosters community instead of seeking maximum profit. "I am not a rich man according to most human reckonings," the man continued. "But I am rich in ability and rich in knowledge, I'm rich in favors and I'm rich in cooperation with others."

The important lesson is that each culture is just one narrow slice of the broad spectrum of human possibility, that our way is not the only human way. If our current civilization now seems bent on self-destruction, its "madness" arises naturally and logically from the modern era's vision of the world and from social and economic institutions that express that vision. While all human societies have possessed and exercised the cultural capacity to shape the world, in the modern era we have pursued power and control—abetted by fossil fuels, science, and industry—with an aggressive intensity that makes our civilization unique.

* * *

Every human society has a shared picture of the world, as well as some sort of guiding story, which together make up its cultural map, or mythology. Typically, however, people are unaware or only partially aware of their own cultural map, because its central beliefs are pervasive and hence as invisible as the air we breathe. Yet, it is these *implicit assumptions*—which seem so natural and true that they are never scrutinized or questioned—that are decisive and dominant in a given culture and era.

The nomadic Gabra in northern Kenya inhabit a landscape permeated with a sense of the sacred and see the world as a place constantly being created and replenished in life's unending cycles. The aim of life, in their view, is *finn*—fertility or plenty. Humans can foster this plenitude of life in many ways—by caring for their animals and for the Earth, cultivating friendships, sharing ideas, singing songs, and telling stories.

Medieval Christians understood life through the framing story of the Fall of Adam, which described the Earth as the foul, polluted realm of Satan and a place of exile for sinful humans. In the High Middle Ages, this Christian vision merged with classical Greek astronomy and philosophy into the Earth-centered world picture of Dante's *Divine Comedy*—a universe in which perfectly ordered, nested spheres ascended from the foulness of Hell at the center of the Earth upward to the incorruptible heavens and finally to the highest, outermost sphere, the throne of God.

One of the oft-repeated myths of the Scientific Revolution, which appears in the writings of such popular science authors as Stephen Jay Gould and Carl Sagan, ascribes the strident opposition to Copernicus's heliocentric system to human arrogance and resistance to being displaced from a privileged spot at the center of the universe. The truth was, in fact, exactly the opposite. In the hierarchy of the spheres, the position of Earth, at the center and farthest from the throne of God, was a sign of human humiliation rather than exaltation. Copernicus's new scheme, which placed the sun at the center of the planetary system, presented a philosophical problem chiefly because it put the fallen Earth into the heavens, which, in the medieval worldview, was the realm of perfection and the eternal. Galileo's observations through his telescope caused further consternation because he reported comets, sunspots, and new stars, all evidence that the heavens were not unchanging or unblemished.

For Dante and his contemporaries, the geocentric universe was not simply a description of the physical world, as it had been for Greek philosophers; in the medieval synthesis, it had become charged with religious symbolism and thus, in situating Earth-dwelling humans between Heaven and Hell, it physically depicted the Christian drama of sin and salvation. The aim of life for those guided by this cultural map was to escape upward from this fallen world.

This doctrine of otherworldliness, which remained the official teaching of the Church throughout the medieval period, could never fully suppress people's abiding interest in this world; a strain of humanism grew stronger as Europe emerged from the centuries of political disorganization, invasions, and insecurity that followed the fall of the Roman Empire. By the time of the Renaissance in the late fifteenth and sixteenth century, humanism was in full flower, and people now surveyed a beautiful, beneficent, nurturing Nature rather than a Satanic realm. In this new atmosphere, the French humanist Michel de Montaigne could rhapsodize: "Whoever contemplates our mother Nature in her full majesty and luster is alone able to value things in their true estimate." Though religion remained a powerful force, people shifted their attention from seeking eternal salvation to pursuing worldly goals, as European civilization focused increasingly on the full development of the individual and sought greater control of natural forces.

The pursuit of knowledge, therefore, took on a new objective. The ancient Greek philosopher Socrates had held that the aim of knowledge is virtue. For medieval Scholastic philosophers of the thirteenth and fourteenth centuries, the purpose of knowledge was to better know God and to understand how things in the world were linked to this transcendent reality. In the early seventeenth century, the prophet of the dawning of the scientific age, Francis Bacon, proclaimed a revolutionary new doctrine: "knowledge is power." Bacon's goal was not intellectual power born of reflection nor insight regarding how to lead the good life. He meant physical power, the kind of power represented in a bulldozer or genetic engineering, power that can move mountains or, as Bacon envisioned in his scientific utopia, *New Atlantis*, create novel organisms—power, in short, to remake nature according to human needs and desires. Bacon therefore proposed a sweeping reform of natural philosophy—the study of nature—for reasons that were at once ethical and religious. His first priority was to improve the quality of life through the proper application of the mechanical arts and science. He promised that this reform would create "a blessed race of Heroes or Supermen who will overcome the immeasurable helplessness and poverty of the human race, which cause it more destruction than all giants, monsters, or tyrants, and will make you peaceful, happy, and secure." At the same time, Bacon believed the progressive advance of knowledge would also slowly reverse the consequences of the Fall, allowing humans to recover the Godlike powers Adam had supposedly enjoyed in Eden. One of his earliest statements of this project, *The Masculine Birth of Time*, carried the subtitle *The Great Instauration [Restoration] of Man's Dominion Over the Universe*. Practical efforts to improve the tools of survival thus became part of a larger redemptive mission, and Bacon, a figure of immense and enduring influence, injected this potent alloy of utility, reforming zeal, and religious hope into the modern outlook.

The guiding story that has propelled the modern era onward toward this desired dominion has been the narrative of progress—the belief that history is a continuous, cumulative record of growing human knowledge, ever expanding power over nature, and improvement in the conditions of human life. This dominant and pervasive faith that history is moving inevitably toward complete human mastery of nature has reigned, as historian J. B. Bury put it, as "the animating and controlling idea" of modern Western civilization. As the belief in a literal Eden faded, the progress narrative has served, in effect, as the modern era's origin myth, providing an explanation of the world and human destiny and the foundation for much of its meaning, key aims, and values. This story of inevitable advance has been the grounding assumption, the fertile soil, supporting the ideas and institutions that have shaped our current civilization: science, technology, industrial capitalism, the imperative of economic growth, the pursuit of ever greater material wealth and comfort, freedom, and individualism. If medieval Europeans were obsessed with sin and salvation, their modern descendants have been obsessed with power and autonomy. They have pursued the intoxicating dream of emancipation—a revolt that began against kings and hierarchy but grew into an ever expanding rejection of constraints of every kind. Over recent centuries, moderns have yearned to be free of tradition, free of society, free as individuals of shared purpose and obligations, free of physical limits, free of history, and ultimately free of Earth.

The Christian story and its successor, the progress myth, have much more in common than one might assume: Both promise an eventual escape from the human condition. The medieval church provided a spiritual answer to the pain and imperfection of life in the hope of a blissful afterlife; the march of progress has offered a *material* answer in a future on an Earth perfected through advances in science and technology. Even the striving for immortality became a literal, scientific pursuit rather than a symbolic and spiritual one. In the early days of the Scientific Revolution, the seventeenth-century Puritan reformer Samuel Hartlib was already dreaming of the day when scientific and medical advances will "preserve men not only from sickness but from death itself." Today, researchers in the field of aging still chase the same dream of a technological fix for mortality. As the Christian story gave way to the modern era's secular myth, Progress may have replaced Providence, but the newer narrative continued to treat all of human history as a unified drama unfolding according to an inevitable logic and infused with redemptive meaning. In providing such reassurance, a refuge from uncertainty, the faith in progress filled some of the same human needs met by traditional religion.

The belief that civilization is advancing toward some heaven on Earth has also provided consolation in the face of traumatic losses that have accompanied the spread of industrial capitalism over the past two centuries. The widespread destruction, the ugliness, the constant upheaval and displacement of communities have often been accepted with a fatalistic shrug in the faith that tomorrow will more than redeem the heartrending losses of today. In this vein, the inventor and automotive pioneer Charles Kettering declared, "The price of progress is trouble, and I don't think the price is too high."

The ozone hole—caused by the CFCs that Thomas Midgley developed at Kettering's behest—proved otherwise. The modern narrative, with its assurance of the inevitability of history, has implied that we are powerless to resist because, in the common refrain, "You can't stop progress." Mary Midgley has aptly summed up the unfortunate consequences: "The belief in inevitable progress can be used and has been used to justify bad changes that were preventable." Perhaps worst of all, this unquestioned faith that tomorrow will arrive new, improved, and more prosperous has served to relieve those living in the present moment from responsibility for the future. The future, the beneficiary of the engine of progress and technological innovation, would take care of itself.

This idea of ongoing, cumulative progress envisioned by Bacon ran into initial resistance. As the worldly approach to life gained ground in the sixteenth and seventeenth centuries, the inherited Renaissance picture of the world proved increasingly at odds with an expanding commercial society and with the desire, shared by early capitalists and the pioneers of the Scientific Revolution, to exploit, manipulate, and transform nature. The reigning image of the cosmos during the Renaissance was of "a living animal . . . a vast organism, everywhere quick and vital, its body, soul, and spirit held tightly together." Those living in the time of Shakespeare and Bacon generally described this thoroughly alive Nature as a bountiful, nurturing mother. People thought of the Earth itself as a human body writ large, an analogy that led some to blame the tumult of earthquakes on planetary flatulence. Leonardo da Vinci, the towering, all-around genius of the Renaissance, likened the constant flow of water down rivers to the sea to the circulation of blood in animals.

The problem with this organic image of nature, as historian Carolyn Merchant detailed in *The Death of Nature*, is that it contained ethical constraints that ran counter to the forces shaping the new era. In a sixteenth-century work on mining titled *De Re Metallica*, the German scholar and scientist Georg Agricola sought to overcome the moral qualms of those who saw mining as a sordid violation of a living female Earth. Without the benefit of mining, he argued, humans would be condemned to living again in caves and digging out roots with their fingernails. It was the debut of an argument that would become familiar in controversies over destructive exploitation. In the spirit of Agricola, bumper stickers today declare that those who oppose nuclear power plants or drilling oil in the Arctic National Wildlife Refuge deserve to "freeze in the dark."

This growing tension between the organic image of nature and the accelerating social, intellectual, and economic changes led in time to a radical revision of the Renaissance world. "The new commercial empires began to demand an ideology that presented Nature only as a material system to be exploited," according to historian of science Peter Bowler. "If people were to feel comfortable when they used the earth for their own selfish ends . . . , Nature had to be despiritualized."

One of the pioneers of the Scientific Revolution, Robert Boyle, was forthright about his desire to banish any reverence for nature: "The veneration wherewith men are imbued for what they call nature has been a discouraging impediment to the empire of man over the inferior creatures of God." The seventeenth-century English chemist and physicist further complained that such attitudes led many to regard the program to conquer nature as "impious" as well as impossible. Boyle was a leading advocate for the revival of atomism, a natural philosophy from classical Greece that held that all of nature and the visible world we see around us is in some sense an illusion. This powerful, unqualified materialism reduced the rich diversity of the world to the meaningless motion of atoms in the void.

By the second half of the seventeenth century, as this atomism combined with the widespread fascination with such complex autonomous machines as clocks, a new image of nature emerged. Mother Nature—a living, active, creative, self-integrating organic whole—vanished as such leading philosophers as the French mathematician René Descartes began to explain the natural world as a lifeless mechanism. Descartes, who also played a key role in the revival of atomism, rejected the randomness of the classical Greek version of this material philosophy, asserting rather that this mechanical order behaved predictably according to laws ordained by God, now characterized as a clockmaker who had set the physical world in motion and then withdrawn from the operation. This notion of mechanism embraced everything; thus he regarded even the human body and animals as no different from a clock or other mechanical "automata." This extreme stance led him to make some preposterous-sounding claims: He denied that animals possess souls, minds, or sensation, writing that "there is nothing in the whole of nature which cannot be explained in terms of purely corporeal causes totally devoid of mind and thought." Descartes further declared that animals do not feel pain, because they lack under-standing; what appears to be pain is really only a mechanical response, which Cartesians likened to the scream of a pipe organ when a key is pressed. Although this bold assertion flies in the face of common sense and the natural empathy most people experience, it helped to justify the practice of experimenting on still-living animals—vivisection—to advance physiological research, which was an exciting frontier of sci-ence in this period.

This image of world as machine is the deep metaphor at the heart of modern culture and, "despite many reformers," Midgley notes, "Descartes still rules." Seeing the world through the metaphor of the machine does not mean that we take it as literally as Descartes and fail to see the difference between a human body and a clock; rather the machine worldview lives on in our culture's long-standing assumption that living things, which may look dazzlingly complex to the eye, are in reality simple. The atomistic vision—which denies that the whole has its own unity and integrity, so it is no greater than the sum of its aggregated parts—has justified reductionism in scientific investigation and the modern notion of the primacy of self, which has fostered an increasingly radical and socially corrosive individualism. In the machine world, the parts are the fundamental reality and priority—whether these parts are "selfish genes" in evolutionary the-ory or the individual in the social-contract theory of liberal democra-cies. Yet from a commonsense perspective, it is clear that society and the culture precede and shape the individual and that an organism is not just a vehicle for the perpetuation of "selfish genes." Our culture reads the world so naturally through the metaphor of the machine, it does not strike us as the least bit bizarre to describe the human body as "a beautiful machine," as a television exercise program did, or to title a popular science book on evolution and ecology *The Machinery of Nature*, or to describe the human brain as a twenty-first-century automaton—a computer with hardware and software.

More than three centuries later, this basic modern premise about the simple, orderly nature of reality persists, and though modern physics has moved beyond this type of materialism, a good part of science still approaches the world with the atomistic, mechanistic assumptions laid out by Descartes, Boyle, and others in the Scientific Revolution. Scientists pursuing the human genome project proceeded on the mechanistic expectation that they would find the "secret of life" by mapping genetic sequences. They assumed that analyzing the parts, the reductionist approach advocated by Descartes, would give them the "program" that makes a person. No doubt the most stunning discovery at the end of this effort has been how little the genetic sequence of DNA reveals about the process that creates an organism. Contrary to the expectations that have driven molecular biology since Watson and Crick described the double helix of DNA in 1953, life's secrets have proved complex rather than simple.

The revolutionary change that launched the modern era's radical cultural experiment involved two distinct steps: first, the demotion of Nature into mindless mechanism; second, the bold elevation of humanity vis-à-vis the larger world. Bacon reflects this immodest view of humans when he begins his *Refutation of Philosophies* with the declaration: "We are agreed, my sons, that you are men. That means, as I think, that you are not animals on hind legs, but mortal gods." The upshot was the creation of a yawning chasm between humans and the rest of life. In this dualistic vision, humans, who appeared to verge on divinity, stand starkly opposed to a Nature reduced to malleable matter. Bacon faults the natural magicians of the Renaissance—alchemists and other scientific forebears who studied nature and conducted experiments in the hopes of directing natural forces toward human ends—for being too restrained in their collaborative approach to nature. With the new technologies, men could have the power "not only to bend nature gently, but to conquer and subdue, even shake her to her foundations." Even though his scientific utopia proposed an essentially mechanistic approach to solving problems by breaking them down into parts, Bacon's writings are full of violent, vivid, sexually charged metaphors in which he often personifies Nature as a recalcitrant woman. Promising that the new science would bring about "the masculine birth of time," he declares, "I am . . . leading to you Nature with all her children to bind her to your service and make her your slave." Guided by this overweening ambition, moderns have pursued an extreme, aggressive, grandiose notion of dominion—Dominion with a capital D.

The image of the world as a machine meshed well with Bacon's program to regain Eden by extending human control over nature. Transforming animals, plants, and natural systems into automata did more than banish bothersome scruples and reverence; it made the enterprise of science and the dream of human control seem possible. Machines, after all, are human creations and are by design under human control. Unlike a living nature, machines do not change in unpredictable ways. If the world is imagined as a giant clock, that suggests it is simple, orderly, predictable, fully comprehensible, and open to manipulation. By taking a machine apart, it is possible to understand fully how it works. The Eden that Bacon aimed to recover bore little resemblance to visions of a peaceable kingdom where the lion lies down with the lamb and humans live in democratic fellowship with the animals. In Bacon's version, the recovery of Eden did not promise a return to some original harmony, but rather the attainment of absolute power. This is Eden as empire rather than a garden—a realm where humans, who are emphatically not a part of nature, would rule as foreign tyrants over alien subjects. Fantasies of omnipotence, autonomy, and human apotheosis have been the fevered "dreams of reason," as René Dubos put it, and they shaped the modern cultural map as much as the "death of Nature."

The indisputable success of Western science over the past three hundred years attests that the machine model of nature has, nevertheless, been a powerful tool, but it is a limited one. Within the classical scientific framework, scientists focused on aspects of nature that were more or less linear in their behavior, orderly phenomena in which effect follows predictably from cause. The very success of this approach "has tended to obscure the fact that real systems almost always turn out to be *non-linear* at some level," as Paul Davies and John Gribbin have noted. "When nonlinearity becomes important, it is no longer possible to proceed by analysis. . . . The very concept of scientific 'analysis' depends on this property of linearity—that understanding the parts of a complex system implies understanding the whole." One sees the limits of conventional analysis in large ecosystem models. Ecologists often have reasonable success with mechanical models of single processes or isolated populations. But when the component models are spliced together, as ecologist Robert Ulanowicz pointed out, a problem arises. These aggregated mechanical constructs are "notoriously poor at predicting behavior." The reason, he argues, is the underlying machine metaphor, which is fundamentally at odds with the nature of large, complex, open systems.

The modern image of world as machine is falling away, undermined by challenges arising within science itself, as well as by real-world events. The living organisms in nature—evolving, nonlinear, open systems that stay alive by taking in energy and resources from the surrounding environment—cannot be adequately understood by means of the simple, universal laws sought by the classical science based on mechanical philosophy. Indeed, even nonliving matter is anything but the passive, inert lumps of stuff imagined by atomistic mechanical philosophy. When matter, such as sand or water, is part of such a system with energy flowing through it, astonishing things begin to happen. Matter begins to organize itself; order and pattern suddenly emerge. As a pot of water comes to a boil on the stove, the molecules in the unorganized pool fall into a circular motion and divide into an orderly series of convection cells that cycle together in a synchronized manner. The circular movement of heated air rising from the hot desert and descending also generates order as it moves the sand particles around, leaving hexagonal patterns like the cells of a beehive in the sand. The pervasive pattern and organization in the nature we see around us—whether in a hurricane or the improbable stability of the Earth's atmosphere—is the dynamic order of far-from-equilibrium systems, order that depends on the constant supply of energy to maintain integrity and structure. This flow of energy is the music that sets chaotic matter dancing, giving birth to life and the beauty of the world.

These far-from-equilibrium systems behave as coherent wholes that can only be understood in their entirety. For this reason, the analytic approach that seeks to understand the whole only by studying individual parts simply does not work, because this whole emerges from the relationship of these parts and thus transcends the mere sum of them. The shift away from atomism, mechanism, and reductionism has been taking place on many fronts: in the study of far-from-equilibrium thermodynamics and self-organizing complexity by chemistry Nobel laureate Ilya Prigogine; in the "ecology of mind" theory set forth by anthropologist and systems theorist Gregory Bateson; in the flowing process physics of David Bohm, which has much in common with ancient Greek philosopher Heraclitus's flux; in Nobel laureate Barbara McClintock's intuitive, empathetic, antireductionist "feeling for the organism" and what might be called "participatory science"; and in the far-from-equilibrium metabolism of Earth described by James Lovelock. Starting from very different perspectives, these various paths of exploration converge on a complex, dynamic, holistic picture of the world.

If the machine world is crumbling, where is the organizing image, the guiding vision to help us at this critical turning point in the relationship between humans and Earth? The planetary era will be shaped by the image of a creative, dynamic, living Earth—a deep metaphor that is archetypal and ancient and at the same time new. This conviction that the Earth is itself alive in some sense has been expressed through the ages in various religious and cultural traditions, including our own as recently as the Renaissance. In *Timaeus*, Plato's dialogue describing the physical nature of the world, the philosopher describes the Earth as "that single great living creature . . . of which all other living creatures, severally and generically, are portions."

Now, however, this perception is returning through the door of modern science. As described earlier, this resurrection of an organic Earth began with James Lovelock's epiphany regarding the Earth's far-from-equilibrium atmosphere. Viewed even from the distance of space, the chemistry of the atmosphere bears witness that Earth is a living planet, while it is just as evident that our sister planet Mars is lifeless. Since then, Lovelock, his collaborator Lynn Margulis, and others have been investigating how the Earth's myriad organisms play an essential role in planetary self-regulation. In this new view, our home planet is more than the "third rock from the sun" with a thin veneer of life—the biosphere—clinging to its crust. Earth *as a whole* is a living system embracing rocks, oceans, atmosphere, and all of Earthly life.

Early on, Lovelock searched for a proper title for his germinating idea and considered such infelicitous but scientifically respectable possibilities as "biocybernetic universal system tendency/homeostasis." Then his friend and neighbor William Golding, a novelist and Nobel laureate, made a fateful suggestion: If the Earth is indeed a living thing, it should have a name. Why not christen this bold notion Gaia, after the ancient Greek goddess of the Earth, the all-mother, the matrix from which all else issues forth? Lovelock did. The decision proved brilliant, radical, problematic, and heretical in the eyes of his scientific peers.

According to Lovelock's now widely recognized Gaia theory, the Earth is an enormous ancient whole, in which organisms and the material environment are tightly coupled—collaborating, coevolving, and constituting a single self-regulating system that has kept the Earth hospitable for life through 3.8 billion years. This post-Darwinian understanding of Earth's history reveals that living creatures have done more than merely "adapt" to conditions on Earth, they have played an active role in creating and maintaining the world we know. The origin of species is but a part of the larger Earth story, the persistence of the self-sustaining enterprise that has carried on through the evolution and extinction of billions of species. This great collective process known as Gaia has served to perpetuate life as a planetary phenomenon. Though 99.9 percent of the perhaps 50 billion species that evolved here are now extinct, life writ large has not only endured through repeated catastrophes, but seized the creative opportunities presented by disaster. Individuals, species, even ecosystems may be fragile and fleeting, but life itself—the great energy-driven, self-organizing process—has proved itself robust to the point of seeming immortal.

From the very beginning, Lovelock's flash of insight about the uniqueness of Earth was more than scientific, and the idea he named Gaia has been more than a single thing. Like all grand organizing ideas, Gaia is not just another scientific hypothesis to be tested, but rather a whole new way of looking at life on Earth. It arises from the kind of imaginative vision that underpins the best of science as well as poetry. "All great theories are expansive," observed evolutionary biologist and historian of science Stephen Jay Gould, "and all notions so rich in scope are underpinned by visions about the nature of things. You may call these visions 'philosophy,' or 'metaphor,' or 'organizing principle,' but one thing they are surely not—they are not simple inductions from observed facts of the natural world."

Over more than three decades, Lovelock has advanced, developed, rethought, refined, and restated his evolving notion of Gaia in scientific publications and in several lively, often eloquent and lyrical books aimed at a general audience as well as fellow scientists. This new vision of Earth proved to have widespread popular appeal beyond the field of science—among those seeking an integrating perspective in a time of global change and those searching for an alternative to the bleak material nature of modernity, as well as among New Age devotees of the resurrected Earth goddess. But Lovelock's grand organizing idea met with opposition and outright rejection in some scientific circles for reasons that had to do both with the unorthodox way he chose to present it and with narrow specialization and conservative ideas in the mainstream science community. Some scientists simply could not get past the name Gaia, with its suspicious aura of myth and reverence—the attitudes toward nature that leading pioneers of the Scientific Revolution like Boyle aimed to eradicate. So they never got to the substance of Lovelock's idea.

Within science circles, Lovelock's Gaian view of Earth got its best reception from those familiar with nonlinear systems, notably physicists, chemists, and engineers. But biologists were another matter. Such leading evolutionary thinkers as Stephen Jay Gould and Richard Dawkins flatly rejected Lovelock's hypothesis. "The Gaia theory thrives on an innate desire, mostly among laypeople, to believe that evolution works for the good of all," Dawkins stated dismissively. "Profoundly erroneous." Lovelock's top-down view of an evolving system is conceptually at odds with reductionist neo-Darwinian theory focused on natural selection acting on individual organisms or genes. If Richard Dawkins insists that enduring life resides in the individual "selfish gene," Lovelock is equally adamant that life is a systemic, planetary-scale phenomenon. The vessel of enduring life in the Gaian view is the planetary process. Orthodox neo-Darwinians could not conceive within their theory how Earth as Gaia could evolve: Because it is unique within the solar system, the mechanism of natural selection could not act upon it. They likewise objected that the Earth system does not meet the definition of a living thing—again according to their theory's definition—because it does not reproduce.

When biologists demanded that Lovelock describe the "mechanism" responsible for planetary regulation, Lovelock sought to show, using a mathematical model known as Daisyworld, that feedback mechanisms of planetary consequence could evolve as organisms furthered their own survival. This line of investigation describes how environmental regulation can arise readily from the fact that living organisms can only grow within a limited range of conditions. If an organism multiplies and alters the environment in a way that makes conditions less favorable for its own growth—for example, by raising the temperature too much—its growth will slow or stop altogether, and as it does, temperatures will moderate. This negative feedback triggered by an organism's physiological limits operates in a way different from Darwinian natural selection and that, Gaia theorists argue, provides a basis for the phenomenon of self-regulation in a living system.

But Daisyworld does not fully explain how regulatory behavior could arise on the planetary scale. Recently, two scientists from a younger generation of British Gaian theorists—climate researcher Richard A. Betts and Earth systems specialist Timothy M. Lenton—proposed that such regulation could evolve through what they call "sequential selection," a mechanism that is "effectively 'trial and error' learning" and one that does not depend on natural selection within a population of competing systems. They argue, moreover, that "the presence of life on Earth, with its capacity for death and mutation, endows the planet with special properties that increase the probability of regulation emerging." There is a sort of selection taking place, but it takes place among a sequence of natural systems over time, which enables a planet to explore different feedback systems and eliminate those that create instability or "anti-regulatory" outcomes, thus "loading the dice in favor of the emergence of self-regulation." Others theorize that over millions of years, natural selection has, however, strengthened the role living organisms play in the chemical cycling that maintains Earth's planetary self-regulation. Organisms that succeed in finding ways to cope with challenges in the immediate environment prosper, multiply, and over time come to regulate large-scale phenomena, such as the production of limestone in the ocean.

More than three decades ago, the eminent medical researcher and essayist Lewis Thomas viewed the question of whether Earth is a living entity through the lens of another discipline, cell biology. He had no trouble accepting that the Earth was alive, but he found the notion that it was like an "organism" unconvincing: "It is too big, too complex, with too many working parts lacking visible connections. The other night, driving through a hilly, wooded part of southern New England, I wondered about this. If not like an organism, what is it like, what is it most like? Then, satisfactorily for that moment, it came to me: it is *most* like a single cell." Thomas went on to explore this arresting metaphor further in a fascinating essay, which describes the Earth's atmosphere as "The World's Biggest Membrane"—one that has the general purpose of "editing the sun." Today Thomas might find the metaphor of organism less problematic. In the intervening decades, advances in Earth systems science have revealed strange and wondrous connections, many of them operating at great distances, that do indeed bind the "working parts" into a single system.

The main obstacle facing Lovelock and his big idea has been the very nature of modern science. "Gaia represents the kind of hypothesis that can only be taken seriously by scientists who have transcended the modern tendency to break everything down into discrete units," judges Peter Bowler, who is the author of a leading history of environmental sciences. "The scientist who studies one of these processes in isolation will inevitably fail to appreciate the overall picture." With specialization and fragmentation into isolated disciplines, scientists are frequently indifferent or actively resistant to integrating big-picture thinking. Stephen Toulmin, a philosopher and historian of science, has lamented that nobody in science has been responsible for looking at the world as a whole. Moreover, this piecemeal approach "deprives us of any standpoint from which to ask fully comprehensive questions, transcending the particular standpoint of any single discipline." Even worse, according to Mary Midgley, in some scientific circles during much of the twentieth century, "the very word 'holistic' has served simply . . . as a term of abuse." The prejudice against Lovelock's idea was so extreme in some quarters that one scientist was warned that he might severely damage or ruin his career if he published an article with the word *Gaia* in the title.

Now, from the perspective of more than three decades, it is clear that science has benefited from Lovelock's holistic provocation, and, though scientists continue to debate myriad specific questions within this framing concept, his insight that life regulates conditions on Earth has gained respectability. Whether the taboo word was mentioned or not, Gaia has helped foster investigation of broad-range interactions and the kind of synthesizing studies that we need to confront a planetary emergency. Stanford University climate scientist Stephen Schneider, who did much to encourage serious discussion of the Gaia hypothesis by organizing two prestigious conferences on the topic, credits Lovelock with helping overcome the hobbling specialization within science. "The Gaia hypothesis is a brilliant organizing principle for bringing together people who don't normally talk to each other, like biologists, geochemists and atmospheric physicists, to ask profound questions about how we got here and how the machinery works." In 1998, scientists and scholars from a variety of disciplines, including the celebrated Harvard University evolutionary biologist E. O. Wilson, met in London to launch the Gaia Society to "promote the study of the earth as interconnected living system," a venture that has since become the Gaia: Earth Systems Science group of the Geological Society of London. Then in 2001, more than 1,500 scientists studying the human impact on planetary systems endorsed the Amsterdam Declaration on Global Change, which recognized that "the Earth System behaves as a single, self-regulating system comprised of physical, chemical, biological, and human components." The Gaian framework today gets serious attention in a synthesizing scientific volume on global change and in textbooks that bring several disciplines to bear on "Earth Systems Science."

As with any fruitful metaphor, the notion of Gaia is a complex one that can be viewed from many illuminating angles. Lovelock's top-down perspective and his emphasis on self-regulation have highlighted the large-scale order and broad stability that have generally kept conditions on Earth within a tolerable range for living organisms. But recent studies of the planet's climate history have revealed unexpected dynamism and volatility within this system—a ground-level view of the roller-coaster vicissitudes experienced by particular species, including our human ancestors, in the struggle to stay alive. While living organisms may, as a general rule, foster stability and help maintain conditions favorable to life, they can also jeopardize that stability. This was apparently the case when photosynthesizing microbes altered the Earth's atmosphere, causing the oxygen holocaust and sending the planet reeling toward the near-death experience of snowball earth. The creativity at the heart of life brings disturbance, sometimes even destruction and death, and disturbance, in turn, opens the door to creativity and evolution. Like the all-mother of Greek mythology or the dancing Shiva, the Hindu god of creation and destruction, contemporary Gaia has two opposing faces. Life's rich complexity embraces contradiction and is rife with paradox.

Gaia's most lasting impact, however, may lie beyond science. In the early 1990s, as he was beginning to make some headway among his scientific peers, Lovelock for a time considered renouncing the name Gaia. Colleagues were telling him that scientists would never accept anything named for a Greek goddess, so he would continue to meet with resistance among his peers. His wavering prompted protests and an impassioned plea in the respected publication *New Scientist* headlined "Gaia, Gaia: Don't Go Away": "Gaia was an idea that everybody could grasp. . . . Gaia challenged natural scientists to believe the whole was far greater than the sum of the parts—a grand unifying theory of Earth. . . . If science cannot find room for a grand vision, if Gaia dare not speak her name in *Nature*, then shame on science. To recant now would be a terrible thing. Don't do it."

In the end, Lovelock stayed true to his intuition that Gaia as metaphor is as vital to his integrating vision of Earth as its framework theory has proved productive for scientific inquiry. Like anthropologist and systems theorist Gregory Bateson, he has found that scientific logic alone is inadequate for describing the natural world in both its complexity and its ultimate unity. Metaphor, according to Bateson, and particularly the extended metaphor of religion, has "made it possible for ordinary people to think at levels of integrated complexity otherwise impossible." For this reason, Bateson argued, "the richest knowledge of the tree includes both myth and botany." Through Gaia as metaphor, it is possible to glimpse the organic unity of the Earth and be awed by our own existence within this rich, complex, and wondrous whole. And in both its scientific and metaphoric aspects, this new view of the Earth provides the foundation for a new cultural map that can guide us in the planetary era.

Earth as Gaia speaks not only to this moment in human history, but to deep human needs that have not been met in the modern era. The metaphor of world as machine described a passive, deterministic, alienating nature, a nature that no one would be eager to claim as kin. The same has been true of some recent versions of evolutionary theory, which, in the spirit of the modern era, have viewed life exclusively through the lens of competition and the metaphor of selfish genes. By contrast, the emerging image of nature as a dynamic, creative, open system, the Gaian story of life and Earth as an ongoing collaboration, are optimistic and attractive. The contingency of an unpredictable nature may be the inevitable price for this rich, creative possibility, but it seems far more credible that humans come from such a nature, that we truly belong to such an Earth.

Through Gaia, the possibility for a new cosmology that embraces nature and humanity in a single common order begins to come into view. In one of his speeches reflecting on this critical historical moment, Vaclav Havel, the playwright and former president of the Czech Republic, described the scientifically rooted notion of Gaia as "inspiration for the renewal of this lost integrity." The Gaia hypothesis reminds us "in modern language, of what we have long projected in our forgotten myths, and of what perhaps has always lain dormant within us as archetypes—that is, the awareness of being anchored in the earth and the universe, the awareness that we are not here alone or for ourselves alone, but are an integral part of a higher, mysterious entity. . . . This forgotten awareness is encoded in all religions. All cultures anticipate it in various forms."

The guiding story for this planetary era will place the origin of our species, our human history, and questions about our existence today within the sweeping narrative that recounts how Earth and its commonwealth of life have together made Gaia—the living system born of creative collaboration, the great process that makes possible our lives and all we see around us. The Gaian story weaves human existence back into this ancient drama, showing how we are rooted in the Earth and the entire cosmic history that preceded it. It tells us that our lineage was born through the dynamism of an evolutionary epic that is still being written. It reveals how our species emerged in response to repeated dramatic shifts in the Earth's climate. It tells us about the trials of tough, courageous, creative ancestors who managed to survive within a contingent nature. It forces us to answer the question "Who am I?" as individuals and as a species in a different way than we have in the modern era.

The individualism at the core of modern identity begins to lose its hard boundaries when one recognizes that the Earth system is a nested hierarchy made up of many levels and sizes of "individuals," each composed of still smaller individuals. "A good case can be made for our nonexistence as entities," Lewis Thomas maintained. "We are shared, rented, occupied." Even our smallest part, the cell, is a merger between two originally separate organisms. The mitochondria, which provide the cell's energy, bear the stamp of separate origin in possessing their own DNA apart from the DNA in the cell nucleus. In the prescient words of the nineteenth-century English novelist Samuel Butler, "Every individual is a compound creature." Our bodies are the home for a vast population of microbes, numbering roughly 100 trillion, including an estimated 500 species of gut bacteria that contribute to proper intestinal development, digestion, and the health of the immune system. In a similar way, each of us inhabits a larger "individual," the Earth system or, in Butler's words, "this huge creature LIFE." The Gaian lens blurs separateness and illuminates connection and relationship. It reveals that we are so embedded in this living commonwealth and Earthly process that it is difficult to determine exactly where any individual begins and ends. "Life did not take over the globe by combat," says Lynn Margulis, "but by networking." Our lives are inseparable from this larger order, and all questions of identity and meaning must be answered in relationship to the whole.

* * *

Armed with the outlines of a new map, we can begin to confront the question of how to live prudently and productively within a nature that is sustaining, dynamic, and sometimes lethal.

The modern approach to the world might be described as "let it rip"—do whatever you want and see what happens. Burn up fossil fuels as fast as possible, pour trillions of pounds of synthetic chemicals over the face of the Earth, flood the land with man-made fertilizers, suck up resources from far and wide to keep the ever vaster industrial enterprise churning. In a nature that does not set booby traps, such behavior might be seen as enterprising and bold. After the appearance of the ozone hole, it is reckless and immoral. The modern era has, indeed, been a radical cultural experiment, and Nature is saying no.

How do we approach the practical business of living, now that we understand how human action engages the global whole?

A *New York Times* headline that appeared just before the World Summit on Sustainable Development in Johannesburg in 2002 proclaimed the establishment answer: "Forget Nature: Even Eden Engineered." The modern project of dominion is still alive and well among those who advocate "planet management" with the help of scientific monitoring. It proceeds in the belief that the modern approach to the world can be sustained if we manage the damaging impacts and steer clear of dangerous thresholds—a path that now requires geo-engineering. Its proponents contend that modern science understands the world well enough to manage this tricky task, although recent experience raises ample doubt. This unapologetically anthropocentric and utilitarian approach to nature rests on the faith that technological innovation and increasing efficiency will make it possible to reduce physical stress significantly while allowing for unabated economic growth. The main appeal of this approach is that it does not require that we adapt ourselves and redesign our civilization in any deep way. Its primary aim, it seems, is to sustain the current system of industrial capitalism and its imperative of exponential growth.

Those who advocate "stewardship" typically have more in common with religious traditions than with the utopian engineering spirit of Francis Bacon. They view the question of the human relationship to nature as an ethical concern more than a utilitarian one. They worry more about whether human conduct is right than whether it is effective. Indeed, as evangelical groups have joined the campaign for action on climate change, they have invoked the Biblical commands for human stewardship of God's creation. Those who hold the ideal of stewardship generally place greater value on the health of natural systems and aim to put an end to destructive exploitation for pragmatic and moral reasons. Proponents of stewardship hope to foster more responsible human behavior, which can certainly bring about improvements on the local or regional scale. But this idea loses traction on the planetary scale. Like planet management, the stewardship notion implies that we are in a position to take charge of nature and it thus mistakes our position vis-à-vis the larger world.

Still others have sought to forsake any whiff of dominion in the human-Earth interaction. The metaphors they have summoned to set forth the new relationship suggest an encounter between equals. The environmental historian Carolyn Merchant has advocated a "partnership" with Earth. Some have sought to capture the dynamic of creative interaction by comparing the new relationship to jazz. Metaphors by their nature highlight some things and obscure others. The emphasis on attentive interaction and respect in these metaphors is valuable and will undoubtedly be fruitful in many contexts. But as with stewardship, the question of scale is critical. The notion of a creative dialectic or partnership between equals fails to capture important aspects of our situation in the planetary era. A successful partnership depends on clear communication and understanding of how the other is likely to respond. Jazz depends on a shared musical language. Neither of these metaphors—jazz or partnership—speaks to the dangerous otherness of nature, especially on the planetary scale, and the fact that it may respond in catastrophic ways we cannot anticipate.

In the final pages of *Silent Spring*, Rachel Carson offers what seem to me the best initial guidelines for making our way in the planetary era. Warning about the "vast forces with which we tamper" and their capacity for "striking back in unexpected ways," she advises caution and humility in our interactions with this responsive and sometimes dangerous nature. The relationship she envisions is one of careful and respectful engagement that is ever aware of the limits of our knowledge and prediction. The aim, now that the fantasy of dominion has ended, should be "to achieve a reasonable accommodation" with a nature that will always defy control.

There is no hope for accommodation on the current path. The quest for economic growth was, in the view of historian J. R. McNeill, "easily the most important idea of the twentieth century," and accelerating economic globalization continues to challenge the physical limits of our ultimately finite planet. This growth makes it difficult, if not impossible, to keep heat-trapping carbon dioxide levels low enough to avoid dangerous climate change. Yet in mainstream efforts to find a workable strategy to prevent CO_2 from rising to catastrophic levels, policy analysts never question the wisdom of continuing exponential economic growth. It is simply taken for granted. Leading economists hold out the vague promise that better technologies will come to the rescue and, in the words of Jeffrey Sachs, "square the circle of economic growth with sustainability," but this remains an untethered faith rather than a demonstrable option. The gross world product, which stood in 2008 at $65 trillion, could, according to economic forecasts, reach $275 trillion by the middle of the twenty-first century. The notion that growth in the current mode can continue much longer, much less indefinitely, is, to quote the conservative British political thinker

John Gray, "a wholly unrealisable fantasy" that jeopardizes the human future.

The time is long past to question this obsolete ideology. There is no avoiding what becomes more obvious day by day—the need to reconsider not simply the means by which we pursue endless growth, but the wisdom of the goal itself. There is considerable disagreement about whether constant growth is necessary to maintain economic equilibrium and whether capitalism can function successfully in a society that is not merely an adjunct to the market economy. Until we confront the clash between exponential economic growth and readily apparent planetary limits, we cannot begin a serious exploration of alternatives to the current system and the tricky problem of how to accomplish a transition out of it. John Gray, former World Bank economist Herman Daly, and others believe that the overall societal goal must become a stationary or steady state for both the human enterprise and, some say, the size of the human population. There are no simple answers to these difficult issues, but there will be no possibility for creative thought if we don't have the common sense and courage to even ask the questions. The taboo around the subject of growth has been so powerful that those who raise these questions can find themselves marginalized in the policy establishment.

The overarching goal for the civilization that succeeds the modern era must be homecoming, the unqualified acceptance of Earth as the true human home and the task of creating a viable, enduring culture within its finitude. In a literal, physical sense, of course, this seems nonsensical: We are already home. But in emotional and psychological ways, modern Western civilization has fostered estrangement and a conviction that human destiny is more than Earthly. If Francis Bacon declared that humans are not animals on their hind legs, but "mortal gods," NASA officials and leading scientists proclaim that humanity's "destiny" is in space. The physicist Stephen Hawking is one of several prominent figures in the scientific community campaigning for space colonization, first on the moon in twenty years and on Mars twenty years after that, on the grounds that humans will become extinct unless they leave Earth. The Alliance to Rescue Civilization, supported by a variety of notable figures involved in space policy, including former astronaut Colonel Buzz Aldrin, advocates a base on the moon that will serve as a repository for DNA samples from all life on Earth and a concise compilation of all human knowledge. Insurance is a reasonable goal, but there is something mad in thinking that humans have a destiny apart from Earth. Now flying the banner of survival, space colonization is just the latest version of escapist dreams of reason, which divert us from practical measures that can be taken in the face of clear and present dangers. What an irony that the most sensible comment in a news story about Hawking's proposal came not from a scientist but from science fiction writer Kim Stanley Robinson, author of a trilogy about future human colonization of Mars: "You want to treat this planet like the only one we have because Mars is poisonous."

If humans are to have any chance at a long-term future, we must give up the persistent and pervasive notion that we do not really belong to this imperfect Earth of mortal creatures. We must abandon the conviction, which also has deep roots in the Western tradition, that we are some sort of special creation, mortal gods, noble beings in exile. We must wake from dangerous dreams of escape from the human condition, of emancipation from Earth. We must reconcile ourselves to the truth that death, suffering, and finitude go with the territory as much as life, joy, and beauty. Ironically, the flight from death has fostered a modern civilization that is profoundly antilife in everything from its image of the world as a machine to the factory farms that treat pigs, chickens, and other animals like machines in the name of profit and efficiency.

The fantasy that humans are not wholly of the Earth and the dream of escape from the human condition and from history have been a part of the Western tradition stretching back millennia to classical Greece and to the Persian prophet Zoroaster. In the modern era, this exceptionalism and escapism became only more pronounced. It will surely be difficult to let them go, but accepting our own mortality and the finitude of Earth is the condition for making our way home. In doing so, we gain kinship, a long overdue understanding of who we really are, and the possibility of fashioning an enduring place for humankind within the living Earth and its ancient family of life. This return to Earth is a journey of self-understanding and a matter of survival.

9

Honest Hope

The way through the world
Is more difficult to find than a way beyond it.
—WALLACE STEVENS

In times of danger, bitter truths serve us better than sweet lies.

The modern era has not taken us where our guiding myth promised. The belief in deliverance through progress has been shattered by developments such as the ozone hole and global warming. Our most rapid progress now may be toward making the planet uninhabitable for many kinds of life, including ourselves.

The decades ahead promise unimaginable loss. Much of the world as we know it—the maple forest of the New England autumn, the coral reefs in tropical waters, the polar bears and countless other plants and animals, low-lying islands and sandy beaches, and perhaps even some of the world's coastal cities—will likely vanish in the lifetime of a child born today. These are not the dark prophecies of environmental apocalypse invoked to scare us into changing our ways, but simply inescapable consequences of the change already set in motion. Shutting off all greenhouse gases today will not stop the warming any more than shutting off the engine can stop a runaway train hurtling down a mountain (though it might switch us onto a less precipitous and calamitous track). The century ahead promises to be a wild trip.

Of all the hurdles that lie ahead, the most formidable may simply be to recognize that the world has changed fundamentally and that we must prepare to meet a future that may bear little resemblance to what we have come to expect. There is no guarantee that the centuries ahead will unfold as some recognizable version of life we live today. The long view of the human journey testifies that no one living on a dynamic, changeable Earth is entitled to expect the secure, comfortable life the rich of the world have come to take for granted. For most of human history, we have not lived in complex societies, in significant part because the climate was far too variable to support agriculture or settled civilizations. Over the past 11,700 years, humans have enjoyed the climatic blessings of the long summer, an interglacial period that has been extraordinarily lengthy and tranquil. This rare interlude in climate history, this special landscape of possibility, has allowed humans over millennia to construct a global civilization. Now, as we face its end, tomorrow's possibilities may be altogether different.

It is hard to envision what the next chapter of the human story will be like, especially if the climate system returns to the variability that has prevailed for long periods in the past. Yet when I look at the beautiful and powerful Chauvet cave paintings executed by an ancestor some 32,000 years ago, I recognize something essential, not only about the past but also about the future: Long before the rise of complex civilizations, we were complexly human. The horse heads drawn in charcoal on the tan rock take my breath away, because they are more than a depiction of a thick-necked, heavy-jawed horse with an erect toothbrush-like mane. I once saw the endangered wild equine from the Mongolian steppe, Przewalski's horse, which looks remarkably like the horses on the cave wall. As I watched it graze, I was captivated by the lines of its square, muscular body and its exquisite coloring—not flamboyant like a tropical bird, but soft, warm, and subtle as the golden shoulder shaded downward into chocolate-stockinged legs. The ears had dark edges reminiscent of Japanese sumi-e brush painting. With shadings and bold lines on rock, the hand wielding the black lump of burnt wood had evoked the definition of its jaw and shoulder muscles, the dark outline of the ears, and the stiff black mane against the warm buckskin coloring of its coat, fully capturing the powerful beauty. The ancient artist had left something of his or her human response to the world that could span culture and millennia and touch me to the quick. We had seen wild horses with the same eyes and heart.

If the Earth reverts to wilder music in the decades hence, I want humans, with or without complex civilization, to remain part of the dance. I wish this not because I think we're exceptional within the Earth's commonwealth of life—though we are, in our way, just as the camouflage genius, the cuttlefish, is in its way—or that Gaia or the cosmos needs us for any reason, but because there is a joy in being alive and part of this ancient drama. I want others to experience this and have their moment on the green, exuberant Earth. I like to imagine that someone will be equally taken by the sight of wild horses 32,000 years from now and moved to capture it in words or images. In the course of this exploration, it has become clear to me that modern civilization is not the measure of humanity. It is not the only or best way of being human. It is not a reason to conclude that the Earth would be better off without our species or that we are doomed by our flawed nature to self-destruct.

The danger now confronting us is that our current global civilization, this radical cultural experiment, will destroy not only itself, but conditions that allow for any form of organized human society. In the worst case, the global change driven by modern industrial civilization could shift the Earth into a new state that is inhospitable to our kind of life. If past history is any guide, the Earth and its great Gaian process will survive the assault of the modern era, just as it has survived the oxygen crisis, asteroid hits, and other shattering catastrophes. Given time, life will rebound and head off in new directions, because, ever creative, this awesome Earth process has demonstrated the ability to transform the bitterest of lemons into sweet lemonade.

In this way, poisonous oxygen became the breath of complex life. I pray we avoid such a worst case, that humans survive the fallout from the modern era and have a chance to write a new chapter of our story.

The only certain thing about the coming century is its immense uncertainty. The great temptation of our time will be the impulse to flee from this uncertainty. Given the black-and-white propensity of Western minds, it will take conscious effort to resist taking refuge either in despair—in the conviction that "it's too late"—or in the alternative, to bask in groundless, sunny optimism that "we'll figure out something, because science always does." I have heard a great deal said about the importance of hope as the human prospect has grown darker, but hope will sustain us only if it is clear-eyed. In reflecting about cultural traps that have made past societies incapable of meeting the challenge of changing circumstance, the anthropologist Paul Bohannan asks, "Have they at least figured out some of the things they should *not* do? Or are they running on blind hope? That kind of hope kills." I don't think we have figured it out. I fear blind hope as much as despair.

The flight from uncertainty into the arms of Providence, whether it is faith in a technological fix, deliverance by the invisible hand of markets, or the apocalyptic belief that human history is approaching its end, relieves us from the responsibility for the future and the obligation to make difficult choices, to act, and to shape the future as best we can for those making their way in the midst of what could be wild and calamitous change.

Human cultures have found ways before to confront the essential uncertainty of existence without taking refuge in escapism. The ancient civilizations of the Near East—the Egyptians, the Mesopotamians, and the Vedic Indians—shared a sense that the cosmos, the order of the world, was always threatened by chaos and always would be. The drama of existence unfolds through the never-ending struggle, but the world nevertheless goes on. This ongoing struggle, these myths tell us, lies at the heart of existence. There is no promise of an easy exit. Similarly, a friend shares stories from her Navaho tribe that counsel against seeking an easy way. One does not expect life to be free of difficulty and sorrow.

"Societies founded on a faith in progress cannot admit the normal unhappiness of human life," observes John Gray, the British historian. "We have been reared on religions and philosophies that deny the experience of tragedy." I think he is right when he concludes: "The good life is not found in dreams of progress, but in coping with tragic contingencies."

In my own experience, the reality of crisis has proved far different from what I had imagined. More than anything, I have been surprised at what I learned about myself. Through two life-threatening illnesses, I discovered that one finds strength when one has to and simply endures what seemed beforehand terrifying and impossible. In an emergency that could have been fatal, I was amazed at how readily I mustered the calm clarity necessary to extricate myself before it was too late. By forcing us to be more than life requires in ordinary times, crisis can temper and deepen.

Those who find themselves in the extraordinary times ahead will also discover what my friends in Poland and Bosnia, who lived through long, brutal wars, found in the midst of the struggle and loss. Such moments of great trial are not only the worst of times, but for many they can also be the very best, because one often experiences life at its most precious, intense, and meaningful. I remember my friend Fikret trying to explain this astonishing paradox during a hike that took us into the hills above Sarajevo to see the positions from which Serbian paramilitaries had hunted civilians in the streets below during one of the longest sieges in history—almost four excruciating and deadly years. Despite the horror of it all, he said, he knew survivors who looked back on the siege as the happiest time of their lives. His statement stopped me in my tracks. How could that be? I demanded.

"Because everything during that time was so clear."

Looking ahead, it is natural to focus on the dangers, but those who will be making their way in this uncertain future will also have unusual opportunities, although these may not be of the kind that one would have chosen wittingly. In the struggle to continue the human journey, they may live lives enlarged by a shared sense of great purpose, leavened by imagination, and enriched by the creativity that survival has always required.

Like our ancestors who managed to survive vicissitude and great hardship over the past 5 million years and who found ways to live creatively in a deeply uncertain world, we must have the fortitude to confront this immense uncertainty and find a path through its dark thickets. As has always been the case, we will be sending our children into this challenging future without guarantees. But if we are both wise and prudent, we will arm them for this dangerous passage with understanding about how our inherited assumptions about ourselves and the world have unleashed this instability, with resilient institutions designed for flexibility and redundancy rather than profit and efficiency, with a recast map of the world that reflects the volatile nature into which they have been born, and with the most precious endowment passed down through countless human generations—knowledge, courage, and honest hope.

Notes

CHAPTER 1.   THE FUTURE HEAD-ON

1   *child born today: Bulletin of the Atomic Scientists*, " 'Doomsday Clock' Moves Two Seconds Closer to Midnight," Jan. 17, 2007, http://www.thebulletin .org/content/media-center/announcements/2007/01/17/doomsday-clock-moves -two-minutes-closer-to-midnight. In 1947, the *Bulletin of the Atomic Scientists*, founded by veterans of the Manhattan Project and devoted to global security issues, began monitoring the peril to civilization by means of a now famous Doomsday Clock on the cover of the magazine. The closer the hands move to midnight, the closer, in its assessment, humans stand to the edge of the abyss. In January 2007, the *Bulletin*'s board and distinguished board of sponsors moved the hands two minutes later—from seven to five minutes before midnight. The *Bulletin* has adjusted the hands seventeen times over the past sixty years, but for the first time, it based a change not only on concerns about nuclear weapons but also on the catastrophic harm unfolding with climate change. The statement explaining the latest adjustment recognizes that "over the next three to four decades climate change could cause irremediable harm to the habitats upon which human societies depend for survival." Business as usual in modern industrial society now poses a threat judged to be in the same league as the apocalyptic threat of mushroom clouds.

2   *"long summer"*: Brian Fagan, *The Long Summer* (New York: Basic Books, 2004). Fagan appears to be the first to have called this extraordinarily long, stable interglacial—known to scientists as the Holocene—"the long summer." I use the term for the past 11,700 years, not for the entire 14,700-year post–Ice Age period, which includes an abrupt climate event known as the Younger Dryas, a freeze lasting 1,200 years.

5   *"faster and nastier"*: Michael McCarthy and David Usborne, "Massive Surge in Disappearance of Arctic Sea Ice Sparks Global Warming," *Independent*, Sept. 15, 2006, http://news.independent.co.uk/environment/article1603667.ece. The story quotes Tom Burke, a former government green advisor and visiting professor at the Imperial College in London.

5   *even catastrophic:* James Lovelock, *The Revenge of Gaia* (London: Penguin Books, 2006). See also James Hansen, "Global Warming Twenty Years Later: Tipping Points Near," briefing to the House Select Committee on Energy Independence and Global Warming, June 23, 2008, in which he warned, "Elements of a 'perfect storm,' a global cataclysm, are assembled."

6   *proceed like a smooth escalator:* James Hansen et al., "Climate Change and Trace Gases," *Philosophical Transactions of the Royal Society,* 365 (online May 18, 2007), 1925–1954. This paper states: "Despite these early warnings about likely future nonlinear rapid response, the IPCC continues, at least implicitly, to assume a linear response to BAU [Business As Usual] forcings. Yet BAU forcings exceed by far any forcings in recent paleoclimate history."

7   *past patterns . . . devastating to agriculture:* William J. Burroughs, *Climate Change in Prehistory* (New York: Cambridge University Press, 2005), 51–57.

9   *deep change:* This distinction between ongoing change and periods of fundamental reorganization is made by David Hackett Fischer, *The Great Wave* (New York: Oxford University Press, 1996).

CHAPTER 2.   THE PLANETARY ERA

12   *growing human burden . . . single, self-regulating Earth:* W. Steffen, et al., *Global Change and the Earth System* (New York: Springer-Verlag, 2005), v, 2.

13   *the interaction of all life, the oceans, the air, the soil, and the rocks:* James Lovelock, *Gaia: A New Look at Life on Earth,* rev. ed. (New York: Oxford University Press, 1979, 1995), x.

13   *four nutrients . . . endless interconnected cycles:* R. Socolow, et al., eds., *Industrial Ecology and Global Change* (Cambridge, UK: Cambridge University Press, 1994), 117.

14   *help regulate climate:* R. J. Charlson, et al., "Oceanic Plankton, Atmospheric Sulfur, Cloud Albedo, and Climate," *Nature* 326 (1987): 655.

14   *the sulfur returns to earth:* M. O. Andreae, "Ocean-Atmosphere Interactions in the Global Biogeochemical Sulfur Cycle," *Marine Chemistry* 30 (1990): 1.

14   *distant connections:* Steffen, *Global Change,* 55.

14   *radical new view of life:* This section relies heavily on James Lovelock's account of this period in his books *Homage to Gaia* (New York: Oxford University Press, 2000), *Healing Gaia* (New York: Crown, 1999), and *Gaia: A New Look at Life on Earth.*

15   *"Life . . . is a verb"*: Lynn Margulis and Dorion Sagan, *What Is Life?* (New York: Simon & Schuster, 1995), 20.

16   *behavior . . . fundamental to life*: Margulis, *What is Life?*, 23. In this section, Margulis and Sagan discuss the theory of Chilean biologists Humberto Maturana and Francisco Varela, which describes life as *autopoiesis* or self-creation, a process in which metabolism is key.

16   *in the deepest sense a living planet*: Lovelock, *Homage to Gaia*, 241, writes that our planet "behaves like something alive." This description remains controversial in the eyes of some scientists, but the Earth system, like a living organism, maintains itself through metabolism.

16   *"life is a necessary and active player"*: Steffen, *Global Change*, ch. 2.

17   *"outside of its normal operating range"*: Steffen, *Global Change*, 81.

17   *"the unifying force of an age"*: Jacques Barzun, *From Dawn to Decadence* (New York: HarperCollins, 2001), 466–67.

18   *"self-sustained economic growth"*: Eric Hobsbawm, *Industry and Empire*, rev. ed. (New York: New Press, 1968, 1999), 12–13.

18   *"more or less automatic"*: W. W. Rostow, "The Take-off into Self-Sustained Growth," *Economic Journal*, 66 (1956): 25–48.

18   *The momentum*: This description is distilled from a variety of historical sources: Hobsbawn, *Industry*; Karl Polanyi, *The Great Transformation*, rev. ed. (Boston: Beacon Press, 1944, 1957); J. R. McNeill, *Something New Under the Sun* (New York: Norton, 2000); Arnulf Grübler, "Industrialization as a Historical Phenomenon," in *Industrial Ecology and Global Change*; David Landes, *The Wealth and Poverty of Nations* (New York: Norton, 1999); William H. McNeill, *The Global Condition: Conquerors, Catastrophes & Community* (Princeton: Princeton University Press, 1992).

18   *"accessory of the economic system"*: Polanyi, *Transformation*, 75.

19   *a mere 500 million*: Stephen Pacala, "Equitable Solutions to Greenhouse Warming: On the Distribution of Wealth, Emissions, and Responsibility Within and Between Nations," presentation at the sustainable global development conference of the International Institute for Applied Systems Analysis, Vienna, Austria, Nov. 14–15, 2007, http://www.iiasa.ac.at/iiasa35/docs/speakers/pacala.html.

19   *rise and fall in tandem*: McNeill, *Something New Under the Sun*, 9–10.

19   *people still provided 70 percent*: Vaclav Smil, *Energy in World History* (Boulder: Westview Press, 1994), 226.

19   *earliest reports of this technology:* Smil, *Energy in World History*, 103.

20   *In Roman Britain:* Richard Cowen, *Geology, History, and People*, ch. 11, "Coal," http://www-geology.ucdavis.edu/~cowen/~GEL115/115CH11coal.html.

20   *helped make more coal available:* Cowen, "Coal."

21   *economic growth and population growth parted ways:* Angus Maddison, *The World Economy: A Millennial Perspective* (Paris: OECD, 2001).

21   *second phase of the Industrial Revolution:* Arnulf Grübler, "Industrialization as a Historical Phenomenon," in *Industrial Ecology and Global Change*, ch. 4.

21   *eightyfold:* based on figures in Smil, *Energy in World History*, 187, and J. R. McNeill, *Something New*, 15. Use increased 38 percent between 1990 and 2005.

21   *sixty-eight-fold:* Angus Maddison, "The West and the Rest in the World Economy: 1000–2030," keynote presentation, Conference of the European Association for Evolutionary Political Economy, Rome, Italy, Nov. 7, 2008, Table 8, p. 14. The *CIA World Factbook* estimates 2007 GWP to be $65.6 trillion.

21   *in a single year:* http://www.indexmundi.com/g/g.aspx?v=65&c=xx&l=en.

21   *eleven times more per person:* Maddison, "The West and the Rest," Table 1a, p. 2.

21   *77 percent . . . remain poor:* Branko Milanovic, *Worlds Apart* (Princeton: Princeton University Press, 2005), 131. See table 10.1.

21   *threat to social stability:* Thomas Homer-Dixon, *The Upside of Down* (Washington, DC: Island Press, 2006), 65.

21   *half of the people:* James Gustave Speth, *Red Sky at Morning* (New Haven: Yale University Press, 2004), 131.

21   *These top 20 percent:* Steffen, *Global Change*, 84.

22   *A generation ago:* Ismail Serageldin, "World Poverty and Hunger—the Challenge for Science," *Science* 296 (2002): 54.

22   *In 1950:* Homer-Dixon, *The Upside*, 65.

22   *inequality overall has grown:* Milanovic, *Worlds Apart*.

22   *"the human race has never had it so good":* Homer-Dixon quotes this remark in *The Upside*, 179.

22   *half of the human transformation:* B. L. Turner, et al., *Earth as Transformed by Human Action: Global and Regional Changes in the Biosphere Over the Past 300 Years* (Cambridge, UK: Cambridge University Press, 1990).

23   *two-page centerfold:* Steffen, *Global Change*, 132–33.

23   *"unique in the entire history of human existence":* Steffen, *Global Change*, 131.

24   *the more complex the society:* Joseph Tainter, *The Collapse of Complex Societies* (Cambridge, UK: Cambridge University Press, 1988, 1990), 91.

24   *"The flow of energy":* Frederick Soddy, *Wealth, Virtual Wealth, and Debt* (London: Allen & Unwin, 1926, 1983), 56.

24   *"prominent economists still fail to give due weight":* Herman Daly, *Beyond Growth* (Boston: Beacon Press, 1996), 185–86.

24   *total demand for energy:* Grübler, *Industrial Ecology,* 64.

24   *the economy has managed to outpace:* Homer-Dixon, *The Upside,* 99.

24   *"would have made much difference":* Smil, *Energy in World History,* 205.

24   *a millionfold increase:* Stephen J. Pyne, *Fire: A Brief History* (Seattle: University of Washington Press, 2001), 155.

25   *"enormous energy flows":* Vaclav Smil, *Energies* (Cambridge: MIT Press, 1999), 5.

25   *eightyfold increase in the use of energy:* Smil, *Energies,* 133.

25   *"Industrial man no longer eats potatoes":* Smil, *Energy in World History,* 188–89. The quotation is from Odum's 1971 book *Environment, Power and Society.*

25   *ten kilocalories:* Dale Allen Pfeiffer, *Eating Fossil Fuels* (Gabriola Island, BC, Canada: New Society Publishers, 2006), 21, citing David Pimentel and Mario Giampietro, "Food, Land, Population, and the U.S. Economy," Carrying Capacity Network, Washington, D.C., Nov. 21, 1994, http://www.dieoff.org/page55.htm.

25   *three kilocalories of fossil energy:* John Ikerd, "Agriculture, After Fossil Energy," paper presented at the Iowa Farmers Union annual conference, Ankeny, IA, Aug. 25–26, 2006, http://www.ssu.missouri.edu/faculty/jikerd/papers/Iowa-FmUnion-Energy.htm.

25   *"who created bread from air":* The German physicist and Nobel laureate Max von Laue used this phrase to describe this innovation in his autobiographical reflections.

25   *"40 percent of the current human population":* Vaclav Smil, *Enriching the Earth* (Cambridge, MA: MIT Press, 2001), xv.

25   *"flamboyant period":* Soddy, *Wealth,* 30.

26   *exhausted at least 40 percent:* U.S. Geological Survey World Petroleum Assessment 2000. See David Strahan, *The Last Oil Shock* (London: John Murray, 2007), 76–77, for a discussion of criticism that the USGS assessment of reserves has been overly optimistic. Less optimistic analysts judge that the world has already consumed roughly half of the total reserves.

26   *not even last through this century:* Strahan, *Last,* 217. See also Strahan, "Coal: Bleak Outlook for the Black Stuff," *New Scientist,* Jan. 19, 2008, http://environment.newscientist.com/channel/earth/energy-fuels/mg19726391.800-coal-bleak-outlook-for-the-black-stuff.html.

26   *"All of this is just an interlude":* Smil, *Energy in World History,* 205.

27   *the use of fire:* Pyne, *Fire,* ch. 3.

27   *Earth at Night:* http://antwrp.gsfc.nasa.gov/apod/ap001127.html.

27   *fish stocks:* Steffen, *Global Change,* 6.

28   *lost to soil depletion:* Pfeiffer, *Eating Fossil Fuels,* 11.

28   *The Global Footprint Network calculates:* "Humanity's Footprint 1961–2003," www.footprintnetwork.org/gfn_sub.php?content = global_footprint.

28   *pushing the global nitrogen cycle:* Robert U. Ayres, William H. Schlesinger, and Robert Socolow, "Human Impacts on the Carbon and Nitrogen Cycles," in *Industrial Ecology,* 153.

29   *creating havoc:* Peter M. Vitousek, et al., "Human Alteration of the Global Nitrogen Cycle: Causes and Consequences," *Issues in Ecology* 1 (1997): 737–50.

29   *doubled the amount of nitrogen:* Steffen, *Global Change,* 194.

29   *oxygen levels are too low:* Cheryl Lyn Dybas, "Dead Zones Spreading in World Oceans," *Bioscience* 55 (2005): 552.

30   *problems as well:* Ayres, et al., "Human Impacts on the Carbon and Nitrogen Cycles," in *Industrial Ecology,* 141.

30   *"alarming and sometimes irreversible":* comment by Klaus Töpfer, former executive director of the United Nations Environment Programme, in Dybas's "Dead Zones" *Bioscience* article.

31   *outside the natural range:* Steffen, *Global Change,* 3–4. See the discussion of the Vostok ice core data.

31   *The most recent data from Antarctica:* Jonathan Amos, "Deep Ice Tells Long Climate Story," *BBC News,* Sept. 4, 2006, http://news.bbc.co.uk/go/pr/fr/-/2/hi/science/nature/5314592.stm.

CHAPTER 3.   LESSONS FROM THE OZONE HOLE

34   *2 billion years ago:* Lynn Margulis dates the ozone layer to the time when ozone began forming in the stratosphere some 2 billion years ago, while others date the ozone layer to the time when atmospheric oxygen and ozone reached more or less current levels.

34   *essential protection:* Jeannie Allen, "Ultraviolet Radiation: How It Affects Life on Earth," *Earth Observatory,* http://earthobservatory.nasa.gov/Library/UVB/.

34   *gaping loss:* http://ozonewatch.gsfc.nasa.gov/monthly/climatology_10 .html. NASA displays monthly maps at this website dating to 1979, which show the sudden appearance of dramatic springtime ozone loss beginning in the early 1980s and continuing to this day.

34   *Beyond the poles:* The Scientific Assessment Panel of the Montreal Protocol on Substances that Deplete the Ozone Layer, "Scientific Assessment of Ozone Depletion: 2006, Executive Summary," 2006, p. 5.

35   *In 1930, Thomas Midgley:* This section on Midgley's career and the history of CFCs and tetraethyl lead relies heavily on the detailed history provided in Seth Cagin and Philip Dray, *Between Earth and Sky* (New York: Pantheon, 1993).

35   *fluorine-based compound:* The full chemical name for CFC-12 is dichlorodifluoromethane.

36   *"a kind of intoxication":* "Chemists in Atlanta," *Time*, Apr. 21, 1930, www .time.com/time/magazine/article/0,9171,739111,00.html.

36   *Midgley himself had been poisoned:* Jamie Lincoln Kitman, "The Secret History of Lead," *Nation*, March 20, 2000, www.thenation.com/doc/20000320/ kitman; and Cagin and Dray, *Between Earth*, 35.

37   *"washed his hands":* New York Herald Tribune, Oct. 31, 1924, quoted in Cagin and Dray, *Between Earth*, 49.

37   *"had more impact":* J. R. McNeill, *Something New Under the Sun* (New York: Norton, 2000), 111.

37   *did not yet fully understand the ozone layer:* This section on the advance of scientific understanding of the ozone layer also relies on Cagin and Dray, *Between Earth*, 134–35.

38   *global monitoring network:* Oxford University Department of Physics, "The Virtual Dobson Room," www.atm.ox.ac.uk/main/AOPP/Dobson.html.

38   *Chapman's historic paper:* Sidney Chapman, "A Theory of Upper-Atmospheric Ozone," *Memoirs of the Royal Meteorological Society* 3 (1930): 103–25.

40   *2.4 billion years ago:* C. Goldblatt, T. M. Lenton, and A. J. Watson, "Bistability of Atmospheric Oxygen and the Great Oxidation," *Nature*, 443 (2006): 683–86.

41   *a major driver:* Peter Ward, "Oxygen—the Breath of Life," *New Scientist*, Apr. 27, 2007, www.newscientist.com/channel/life/mg19426012.000-oxygen—the -breath-of-life.html.

42   *photosynthesis breakthrough of:* Lynn Margulis and Dorion Sagan, *Microcosmos* (Berkeley: University of California Press, 1997), 70. My account of early life and microbial evolution relies in large part on *Microcosmos* and other writing

by Margulis and coauthors as well as on Marcello Barbieri's account of the evolution of the Earth's fundamental cycles in *The Semantic Theory of Evolution* (New York: Harwood Academic Publishers, 1985).

43   *far more hospitable:* Kenneth Chang, "A New Picture of the Early Earth," *New York Times,* Dec. 2, 2008, www.nytimes.com/2008/12/02/science/02eart.html.

43   *Danish scientists:* Paul Rincon, "Oldest evidence of photosynthesis," *BBC News,* Dec. 17, 2003, http://news.bbc.co.uk/2/hi/science/nature/3321819 .stm. See also Minik T. Rosing and Robert Frei, "U-rich Archaean sea-floor sediments from Greenland—indications of >3700 Ma oxygenic photosynthesis," *Earth and Planetary Science Letters* 217 (2004): 237–44.

44   *"oxygen holocaust":* Nick Lane challenges Margulis's narrative. For an alternative theory of oxygen's role in the history of life, see his book *Oxygen: The Molecule That Made the World* (New York: Oxford University Press, 2002). Lane disputes the claim that Earth suffered an "oxygen holocaust" and maintains that early life had evolved defenses against oxygen by this time.

44   *a perilous period:* Paul F. Hoffman and Daniel P. Schrag, "Snowball Earth," *Scientific American,* Jan. 2000, 68–75. See also Andrew H. Knoll's *Life on a Young Planet* (Princeton: Princeton University Press, 2003), 206–15.

44   *One leading theorist:* Robert E. Kopp, Joseph L. Kirschvink, Isaac A. Hilburn, and Cody Z. Nash, "The Paleoproterozoic Snowball Earth: A Climate Disaster Triggered by the Evolution of Oxygenic Photosynthesis," *Proceedings of the National Academy of Science* 102 (2005): 11, 131–36, www.pnas.org/cgi/doi/10 .1073/pnas.0504878102.

45   *The answer to the puzzle:* James F. Kasting, "When Methane Made Climate," *Scientific American,* July 2004, 78–85.

45   *only a fraction of 1 percent:* Kristin Leutwyler, "The First Ice Age," *Scientific American,* Nov. 1, 1999, www.sciam.com/article.cfm?id=the-first-ice-age. This figure is from a presentation by James Kasting given at a Geological Society of America meeting in Denver.

45   *100,000 years:* This figure is from Kopp, et. al., "The Paleoproterozoic Snowball Earth."

46   *one of Earth's worst climate disasters:* Kopp, et al. See also Kirschvink's comments in "Bacteria Probably Caused Worst Snowball Earth Episode," www .scienceblog.com/cms/node/8694.

47   *exhaust all the available carbon dioxide:* Knoll, *Life on a Young Planet,* 21.

47   *Establishing these grand cycles:* Barbieri, *Semantic Theory,* 35.

48 *"The urge to form partnerships"*: Lewis Thomas, "On the Uncertainty of Science," *Phi Beta Kappa Key Reporter* 46 (Autumn 1980), reprinted in Philip Appleman, ed., *Darwin* (New York: Norton, 1970), 306.

48 *"but only a few"*: Barbieri, *Semantic Theory*, 155.

48 *"green microbial scum"*: comment by David Catling, who is part of a NASA research team quoted in a NASA Ames Research Center news release, "NASA Scientists Propose New Theory of Earth's Early Evolution," Aug. 3, 2001.

48 *pushed evolution into overdrive*: Hoffman and Schrag, "Snowball Earth," 70.

49 *99 percent of species*: David M. Raup, *Extinction: Bad Genes or Bad Luck?* (New York: Norton, 1991), 4.

49 *far less stable*: Leutwyler, "The First Ice Age," www.sciam.com/article .cfm?id = the-first-ice-age, quoting James Kasting.

49 *the dense summer haze*: This section is based primarily on James Lovelock's account in *Homage to Gaia* (New York: Oxford University Press, 2000), ch. 8.

50 *commercial use of Midgley's CFCs*: Cagin and Dray, *Between Earth*, 66–67.

50 *Midgley had a meeting*: Cagin and Dray, 68–69.

50 *Atmospheric Cabinet*: This first home air conditioner is described in the history of the Carrier Corporation at its website, http://english.carrier.com.cn/about/about_history.asp?navmenu = 5#.

51 *an entirely new use*: DuPont company history of refrigerants, http://refrigerants.dupont.com/Suva/en_US/about/history/history_1940.html.

51 *five victims for every one*: John Miller Jr., *Guadalcanal: The First Offensive* (Minnetonka, MN: National Historical Society, 1949, repr. 1993), excerpt of Chapter 9 available at http://ibiblio.org/hyperwar/USA/USA-P-Guadalcanal/USA-P-Guadalcanal-9.html.

51 *"atomic bomb of the insect world"*: Cagin and Dray, *Between Earth*, 74–78.

51 *40 million*: Dupont cites this figure in its website history of refrigerants.

52 *rhapsodized about the virtues of Freon*: Charles Grutzner Jr., "20 Degrees Cooler Inside," *New York Times Magazine*, June 17, 1945, SM10.

52 *CFC production in the postwar years*: "Global CFC Production 1950–1992," graph by Phillipe Rekacewicz, UNEP/GRID-Arendal, United Nations Environment Programme, http://maps.grida.no/go/graphic/global_cfc_production.

52 *Armed with his electron capture detector*: This section relies on Lovelock's *Homage to Gaia* and his paper "Atmospheric Fluorine Compounds as Indicators of Air Movement," *Nature* 230 (1971): 379.

54   *a lasting legacy:* Sharon Roan, *Ozone Crisis* (New York: Wiley, 1989), 15, details the Department of Transportation's initiation of the multimillion-dollar study to assess the threat of supersonic aircraft, the Climatic Impact Assessment Program.

55   *a long-range concern:* Lovelock, *Homage*, 216.

55   *concluded in 1974:* Mario Molina and F. Sherwood Rowland, "The CFC-Ozone Puzzle: Environmental Science in the Global Arena," John Chaffee Memorial Lecture on Science and the Environment, National Academy of Sciences, Washington, D.C., Dec. 7, 2000, www.ncseonline.org/NCSEconference/ 2000conference/Chafee/.

56   *"very bad manners":* Molina and Rowland, "The CFC-Ozone Puzzle."

56   *What happens to the free chlorine?:* This section relies on Cagin and Dray, Roan, and Rowland and Molina's John Chaffee memorial lecture.

57   *"global environmental problem":* Sherwood Rowland, "Nobel Lecture in Chemistry," award ceremony, Stockholm, Sweden, Dec. 8, 1995, http://nobelprize .org/nobel_prizes/chemistry/laureates/1995/rowland-lecture.html.

57   *a seminal article:* This section on the debate following the publication of the *Nature* paper and the series of assessments before the ozone hole appeared relies on Roan's account.

57   *"fluctuated wildly":* Roan, 89.

58   *"Ozone: The Crisis That Wasn't":* Roan, 111.

58   *40 percent of the ozone layer:* J. C. Farman, B. G. Gardiner, and J. D. Shanklin, "Large Losses of Total Ozone in Antarctica Reveal Seasonal ClO_x/NO_x Interaction," *Nature* 315 (1985): 207–10.

59   *"preposterous and alarmist":* Cagin and Dray, 280, 286.

59   *prevailing assumption:* Maureen Christie, "Data Collection and the Ozone Hole: Too Much of a Good Thing?" *Proceedings of the International Commission on the History of Meteorology* 1 (2004), 99–105. See also John Gribbin, *The Hole in the Sky* (New York: Bantam Books, 1988), 110–12.

60   *to accelerate the phaseout:* The phaseout schedule for HCFCs is available at www.state.gov/r/pa/prs/ps/2007/sep/92598.htm.

61   *"surprisingly rapid changes":* "Scientists Say Antarctic Ice Sheet is Thinning," Reuters, Mar. 30, 2007, www.planetark.com/dailynewsstory.cfm/newsid/ 41175/story.htm.

61   *Robert A. Millikan:* Robert A. Millikan, "Alleged Sins of Science," *Scribner's Magazine* 87, no. 2 (Feb. 1930): 121.

62   *proceeded on faith:* Robert L. Sinsheimer, "The Presumptions of Sci-

ence," in *Limits of Scientific Inquiry*, Gerald Holton and Robert S. Morison, eds., (New York: Norton, 1979), 24–25.

63   *"extremely lucky"*: Paul J. Crutzen, "My Life with O_3, NO_x and Other YZO_xs," Nobel lecture, award ceremony, Stockholm, Sweden, Dec. 8, 1995, http://nobelprize.org/nobel_prizes/chemistry/laureates/1995/crutzen-lecture.html.

64   *. Unpredictability arises:* Gilbert C. Gallopin, et al., "Science for the 21st century: From Social Contract to the Scientific Core," *International Social Science Journal* 168 (2001): 220–29. Also see Daniel R. Sarewitz, et al., eds., *Prediction: Science, Decision Making, and the Future of Nature* (Washington, D.C: Island Press, 2000).

CHAPTER 4.   THE RETURN OF NATURE

66   *swift changes in the realms of ice:* James Hansen, et al., "Climate Change and Trace Gases," *Philosophical Transactions of the Royal Society* 365 (2007): 1925–54, www.journals.royalsoc.ac.uk/content/l3h462k7p4068780/?p = 47a7e0c7b654 42c18300d66e2b0f4b5e&pi = 17.

66   *inescapable by 2030:* "Stern Review on the Economics of Climate Change," report to HM Treasury, United Kingdom, Oct. 30, 2006, 2. A PDF version is available at www.hm-treasury.gov.uk/sternreview_index.htm.

67   *"at the planetary controls":* James Gustave Speth, "A New Green Regime," *Environment: Yale*, Fall 2002, 2. Others have employed similar metaphors to emphasize human dominance and responsibility. In an influential paper in *Science*, four leading researchers—Peter Vitousek, Harold Mooney, Jane Lubchenco, and Jerry Mellilo—concluded a summary of human domination by stating: "In a very real sense the world is in our hands—and how we handle it will determine its composition and dynamics, and our fate." See Peter Vitousek, et al, "Human Domination of Earth's Ecosystems," *Science* 227 (1997): 494–99. In a later work, Speth chooses a different metaphor, an uncontrolled "planetary experiment," which points to a different aspect of our predicament: the risks attending our intrusion into planetary systems and our inability to predict or control the consequences. See James Gustave Speth, *Red Sky at Morning*, (New Haven: Yale University Press, 2004).

70   *some economists have long argued:* Part 2, chapter 6 of the "Stern Review" contains an extensive critique of assumptions, omissions, and limitations of Assessment Impact Models. See also: Stephen H. Schneider and Kristin Kuntz-Duriseti, "Uncertainty and Climate Change Policy," in *Climate Change Policy: A Survey*, Stephen H. Schneider, Armin Rosencranz, and John O. Niles, eds. (Washington, D.C.: Island Press, 2002), 53–88.

71 *"optimistically smooth"*: Will Steffen, et al., "Abrupt Changes: The Achilles' Heels of the Earth System," *Environment* 46 (2004): 9–20.

71 *"surprise free"*: Schneider and Kuntz-Duriseti, "Uncertainty and Climate Change Policy," in *Climate Change Policy: A Survey*, 58.

71 *"the most uncertain"*: "Stern Review," 143.

71 *extremely optimistic*: Schneider and Kuntz-Duriseti, 64–65.

72 *force 1 billion people*: John Vidal, "Climate Change to Force Mass Migration," *Guardian*, May 14, 2007, www.guardian.co.uk/environment/2007/may/14/climatechange.climatechangeenvironment. This estimate comes from a study by the development charity Christian Aid.

72 *catastrophe is defined*: "Stern Review," 153.

73 *won't be able to adapt*: Will Steffen, "Surviving the Anthropocene: The Great Challenges of the 21st Century," presentation given at the Stockholm Resilience Center, Oct. 30, 2007, www.stockholmresilience.org/seminarandevents/seminarandeventvideos/willsteffenthegreatchallengesofthe21stcentury.5.aeea46911a3127427980003219.html.

73 *"a false sense of security"*: Fred Pearce made this comment while answering questions after a talk about his book *With Speed and Violence* at the Cambridge Forum, Cambridge, MA, May 2, 2007.

73 *exaggerated the economic costs*: Martin Wolf, "In Spite of Economic Skeptics, It Is Worth Reducing Climate Risks," *Financial Times*, Feb. 7, 2007, 15. See also Robert O. Mendelsohn, "A Critique of the Stern Report," *Regulation*, Winter 2006–2007, 42–46.

73 *One notable exception*: Martin L. Weitzman, "On Modeling and Interpreting the Economics of Catastrophic Climate Change," 2008, www.economics.harvard.edu/faculty/Weitzman/papers.html. The paper has been submitted to the *Review of Economics and Statistics*.

73 *A survey done in the early 1990s*: W. D. Nordhaus, "Expert Opinion on Climatic Change," *American Scientist* 82 (1994): 45–52, cited in Schneider and Kuntz-Duriseti, 69.

74 *what is taken for granted*: Herman E. Daly, *Beyond Growth* (Boston: Beacon Press, 1996), 6–7.

75 *"get along without natural resources"*: Robert Solow, "The Economics of Resources, or the Resources of Economics," *American Economic Review* 64 (1974): 1–14.

75 *Stern report exaggerates the damage*: Gerard Wynn, "Economists on Climate Change: Do We Really Care?" Reuters, Sept. 9, 2006, www.planetark.org/avantgo/dailynewsstory.cfm?newsid=38268.

76   *"damage risks are bigger"*: Fiona Harvey and Jim Pickard, "Stern Takes Bleaker View on Warming," *Financial Times*, Apr. 16, 2008, www.ft.com/cms/s/0/d3e78456-0bde-11dd-9840-0000779fd2ac.html.

76   *civilization will collapse:* Steffen, "Surviving the Anthropocene."

76   *law-giving God:* Joseph Needham, *The Grand Titration: Science and Society in East and West* (London: Allen & Unwin, 1969), 303.

76   *the turn in sixth-century:* F. M. Cornford, *From Religion to Philosophy: A Study in the Origins of Western Speculation* (New York: Harper & Bro., 1912; repr. 1957), 158–59.

76   *first solar eclipse predicted beforehand:* Patricia O'Grady, "Thales of Miletus," *Internet Encyclopedia of Philosophy* (2006), www.iep.utm.edu/t/thales.htm#SH8a.

77   *lawful gradualness:* Arthur O. Lovejoy, *The Great Chain of Being* (New York: Harper & Row, 1936, repr. 1960), 327–29. Lovejoy explains that the requirement that "there are no sudden 'leaps' in nature" is an essential part of a "faith in the rationality of the world we live in," which for a score of centuries guided and animated Western philosophy and science. "In so far as the world was conceived in this fashion, it seemed a coherent, luminous, intellectually secure and dependable world. . . ."

77   *implied supernatural origin:* The nineteenth-century debate over gradual versus abrupt change in nature and its relation to religion was complex. Georges Cuvier, the brilliant French scientist who founded the study of vertebrate paleontology, established the fact of extinction based on the fossil record and theorized that the Earth had experienced a series of sudden and violent "revolutions," which had wiped out many species. Cuvier read no divine causation into these events, but an English geologist and Anglican priest, Rev. William Buckland, took up Cuvier's idea and argued that the last catastrophe had been Noah's flood. The geologist Charles Lyell, who began his career as an adherent of catastrophism, later rejected Buckland's attempt to link the Bible with geology and developed the countervailing theory of uniformitarianism. Lyell had a significant influence on Darwin's thinking. For a discussion of the debate about the rate of change, see Peter J. Bowler, *The Norton History of the Environmental Sciences* (New York: Norton, 1992), 237–47.

77   *things of natural origin:* Howard E. Gruber and Paul H. Barrett, *Darwin on Man: A Psychological Study in Scientific Creativity* (New York: Dutton, 1974), 126.

77   *how the evolution:* P. Z. Myers, review of Stephen Jay Gould's *Punctuated Equilibrium*, in *New Scientist*, May 12, 2007, www.newscientist.com/channel/opinion/mg19426032.100-ipunctuated-equilibriumi-by-stephen-jay-gould.html.

This author describes how creationists invoke discontinuities in the fossil record as evidence of divine creation.

77   *"completed the revolution"*: J. R. Ravetz, *The Merger of Knowledge and Power* (London: Mansell Publishing, 1996), 107.

78   *to believe the contrary:* Richard Levins and Richard Lewontin, *The Dialectical Biologist* (Cambridge, MA: Harvard University Press, 1985), 38.

78   *IPCC's projections and scenarios:* Fred Pearce, *With Speed and Violence* (Boston: Beacon Press, 2007), xxiv. See this for an account of a scientific meeting on "dangerous climate change" convened in 2005 by the British government and on the distinction drawn between gradual change reflected in typical IPCC climate models and abrupt, nonlinear change involving thresholds, which is not reflected in standard models. The discussion at this Exeter meeting distinguished these as Type I (gradual) change and Type II (abrupt) change.

78   *slow, steady procession:* Spencer Weart, *The Discovery of Global Warming* (Cambridge, MA: Harvard University Press, 2003), 9.

79   *"even the best scientific data":* Spencer Weart, "The Discovery of Rapid Climate Change," *Physics Today*, Aug. 2003, 30, www.physicstoday.org/vol-56/iss -8/p30.html.

80   *rapid climate change is normal:* Paul Andrew Mayewski and Frank White, *The Ice Chronicles* (Hanover, NH: University Press of New England, 2002), ch. 3.

81   *this mild time might have lasted:* Wallace S. Broecker and Robert Kunzig, *Fixing Climate* (New York: Hill and Wang, 2008), 63.

82   *saved us from the onset:* Hazel Muir, "The Ice Age That Never Was," *New Scientist*, Sept. 30, 2008, 32–36, http://environment.newscientist.com/channel/ earth/mg19926721.600-the-ice-age-that-never-was.html.

82   *"any time soon":* James W. C. White, "Do I Hear a Million?" *Science* 304 (2004): 609–10.

82   *"sleeping monsters":* Pearce, *With Speed*, xxiv. Pearce attributes this vivid language to Chris Rapley, director of the British Antarctic Survey.

83   *the great mythical hero Beowulf:* The quotations are by Seamus Heaney, *Beowulf: A New Verse Translation* (New York: Farrar, Straus and Giroux, 2000).

84   *"unequivocal evidence":* Will Steffen, et. al., "Abrupt Changes: The Achilles' Heels of the Earth System," *Environment*, Apr. 2004, 10.

84   *as much as 18 degrees F:* Steffen, et al., "Abrupt," 10.

84   *"devastate modern civilizations":* Steffen, et. al., "Abrupt," 10.

84   *8,200 years ago:* Richard B. Alley, *The Two-Mile Time Machine* (Princeton: Princeton University Press, 2007), 182.

85   *Wally Broecker had long theorized:* Broecker and Kunzig, *Fixing Climate*, 128–29, describe the new evidence that the emptying of the great glacial meltwater lakes did not occur at the right time to make this the trigger for the conveyorbelt shutdown.

85   *Dorthe Dahl-Jensen:* "Greenland Ice Cores Show Drastic Climate Change Near End of Ice Age," *Terra Daily*, July 14, 2008, www.terradaily.com/reports/Greenland_Ice_Cores_Show_Drastic_Climate_Change_Near_End_Of_Ice_Age_999.html.

86   *paper in Science:* Jørgen Peder Steffensen, et al., "High-Resolution Greenland Ice Core Data Show Abrupt Climate Change Happens in a Few Years," *Science* 321 (2008): 680–83.

86   *James White:* "Greenland Ice Cores," in *Terra Daily* and personal communication.

86   *profound consequences for agriculture:* W. S. Broecker, "What If the Conveyor Were to Shut Down? Reflections on a Possible Outcome of the Great Global Experiment," *GSA Today* 9 (1999): 1–7.

87   *If the climate were to shift back:* William J. Burroughs, *Climate Change in Prehistory* (New York: Cambridge University Press, 2005), 101–2.

87   *faster than the model-based projections:* Stefan Rahmstorf, et al., "Recent Climate Observations Compared to Projections," *Science* 316 (2007): 709.

88   *heading for a catastrophic disintegration:* Michael McCarthy, "Dramatic Change in West Antarctic Ice Could Produce 16ft Rise in Sea Levels," *Independent*, Feb. 2, 2005, http://environment.independent.co.uk/article10874.ece.

88   *"We didn't know the process":* Richard A. Kerr, "A Worrying Trend of Less Ice, Higher Seas," *Science* 311 (2006): 1698–1701. The quoted statement was made by Robert Bindschadler of NASA's Goddard Space Flight Center.

88   *hit a threshold:* James Hansen, "Climate Change: On the Edge," *Independent*, Feb. 17, 2006, http://news.independent.co.uk/environment/article345926.ece.

88   *82 feet higher:* Hansen, "Climate Change: On the Edge."

89   *a half billion people inland:* Kerr, "A Worrying Trend," 1698.

89   *a line-by-line analysis:* David Wasdell, "Political Corruption and the IPCC Report?" 2007, www.meridian.org.uk/Resources/Global%20Dynamics/IPCC/index.htm. See also Fred Pearce's articles "Climate Report 'Was Watered Down,'" *New*

Scientist, Mar. 8, 2007, 10, http://environment.newscientist.com/article/mg19325943
.900;jsessionid=JAJBGEDAGANG and "What the IPCC Didn't Tell Us," *New
Scientist*, Feb. 9, 2007, 6–8, http://environment.newscientist.com/channel/earth/
mg19325903.800-climate-change-what-the-ipcc-didn't-tell-us.html.

89   *even 18 feet:* Pearce, *With Speed*, 52, 58.

90   *as much as four times greater:* Bruce A. Wielicki, et al., "Evidence for
Large Decadal Variability in the Tropical Mean Radiative Budget," *Science* 295
(2002): 841–44. Wielicki's comments on the greater importance of clouds come
from Pearce, *With Speed*, 108.

90   *Catastrophic warming:* D. A. Stainforth, et al., "Uncertainty in Predic-
tions of the Climate Response to Rising Levels of Greenhouse Gases," *Nature*
433 (2005): 403–6.

90   *not taken sufficient account:* Richard Black, "Global Warming Risk 'Much
Higher,'" *BBC News*, May 23, 2006, http://news.bbc.co.uk/go/pr/fr/-/1/hi/sci/
tech/5006970.stm.

91   *only half as much as it did:* Sylvia Westall, "Oceans Absorbing Less CO_2
May Have 1,500 Year Impact," Reuters, Apr. 17, 2008, www.planetark.com/
dailynewsstory.cfm?newsid=48010&newsdate=17-Apr-2008.

91   *a study looking at the Southern Ocean:* Deborah Zabarenko, "Southern
Ocean Saturated with Carbon Dioxide—Study," Reuters, May 18, 2007, www.
planetark.com/dailynewsstory.cfm?newsid=41988&newsdate=18-May-2007.

93   *three times the rate anticipated:* "Melting Ice," interview with Mark Ser-
reze of the National Snow and Ice Data Center in Boulder, Colorado, *Living on
Earth*, National Public Radio, week of May 11, 2007 (air date), www.loe.org/
shows/shows.htm?programID=07-P13-00019#feature1.

93   *under way in Siberia:* Ned Stafford, "Melting Lakes in Siberia Emit
Greenhouse Gas," *Nature*, published online Sept. 6, 2006, www.nature.com/news/
2006/060906/full/news060904-10.html. See also K. M. Walter, et al., "Methane
Bubbling from Siberian Thaw Lakes as a Positive Feedback to Climate Warming,"
Nature 443 (2006): 71–75.

93   *1,600 billion tons:* Fred Pearce, "Arctic Meltdown Is a Threat to Humanity," *New Scientist*, March 25, 2009, 32–36.

CHAPTER 5.   A STORMWORTHY LINEAGE

97   *Origin stories:* Mircea Eliade, *Myth and Reality* (New York: Harper &
Row, 1963, 1975) and Leszek Kołakowski, *The Presence of Myth* (Chicago: University of Chicago Press, 1989).

98   *evolved into a narrative of progress:* Carolyn Merchant, *Reinventing Eden* (New York: Routledge, 2003), 11–22.

98   *such popularizers:* Mary Midgley, *Evolution as Religion* (New York: Methuen, 1985), 6.

98   *"from gas to genius":* James R. Moore, *The Post-Darwinian Controversies* (Cambridge, UK: Cambridge University Press, 1979), 167, cited in Midgley, *Evolution*, 6. Spencer's disciple was Edward Clodd.

98   *fundamentally at odds with nature:* George Perkins Marsh, *Man and Nature; or, Physical Geography as Modified by Human Action,* David Lowenthal, ed. (Cambridge, MA: Harvard University Press, 1864, Belknap edition 1964).

99   *"a sweet and wild garden":* Bill McKibben, *The End of Nature* (New York: Random House, 1989), 91.

99   *contrary to either version:* I am deeply indebted in this chapter, which looks at the influence of climatic variability on human evolution, to Rick Potts's seminal work *Humanity's Descent* (New York: Morrow, 1996).

101   *repeated evolutionary experiments:* Ian Tattersall, "Once We Were Not Alone," *Scientific American,* Jan. 2000, 56–62.

101   *hominin species:* There has been a recent debate whether to call bipedal apes *hominins* or *hominids,* which is discussed by Lee R. Berger, "Viewpoint: Is It Time to Revise the System of Scientific Naming?" *National Geographic News,* Dec. 4. 2001, http://news.nationalgeographic.com/news/2001/12/1204_hominin _id.html. I will use the newer term, *hominin,* in the text and retain hominid in places where it appears in quotations.

103   *Potts walked up and down:* Ruth Osterweis Selig, "Human Origins: One Man's Search for the Causes in Time," *Anthronotes* 22 (Spring/Summer 1999), http://anthropology.si.edu/outreach/anthnote/Summer99/anthnote.html.

104   *two opposing themes:* Rick Potts, "Complexity and Adaptability in Human Evolution," in *Probing Human Origins,* M. Goodman and A. S. Moffat, eds. (Cambridge, MA: American Academy of Arts and Sciences, 2002), 33–57.

105   *more flexible types:* Rick Potts, "Evolution and Climate Variability," *Science* 273 (1996): 922–23.

105   *similar thinking ability:* "Neanderthal," BBC *Horizon* program, Feb. 10, 2005. The transcript is available at www.bbc.co.uk/sn/tvradio/programmes/ horizon/neanderthal_trans.shtml.

105   *capable of innovation:* Rowan Hooper, "Neanderthals Bid for Human Status," *New Scientist,* June 13, 2007, 12, www.newscientist.com/channel/being-human/mg19426085.400-neanderthals-bid-for-human-status.html.

106 *explanation lies in ecological:* Clive Finlayson, *Neanderthals and Modern Humans* (New York: Cambridge University Press, 2004).

107 *coldest, driest, harshest:* Francisco J. Jiménez-Espejo, et al., "Climate Forcing and Neanderthal Extinction in Southern Iberia: Insight from Multiproxy Marine Record," *Quaternary Science Reviews* 26 (2007): 836–52.

107 *Homo erectus:* Those who study human origins are still debating whether *Homo erectus*, an extremely successful species that endured for 1.5 million years, was a direct human ancestor. Increasingly, however, evidence weighs in favor of the theory that modern humans evolved in Africa some 200,000 years ago from a species similar to *Homo erectus* and then began to migrate and colonize other regions. Why *Homo erectus* became extinct or whether our species played any role in its disappearance through replacement or limited interbreeding is unclear, especially in light of new studies in Java which suggest that *Homo erectus* populations may have persisted there until as recently as 27,000 years ago—long after the evolution of our kind. See Ian Tattersall, "Homo erectus: Current Debates," *Microsoft Encarta Online Encyclopedia*, 2008, http://encarta.msn.com/encyclopedia_761586359/homo_erectus.html.

108 *developmental disorder:* Recent studies have detailed structural similarities between the Flores bones and those of an early hominin called *Australopithecus afarensis*, the most famous example of which is the fossil discovered in Ethiopia in 1974 known as Lucy. If the hobbits are indeed a new species, then some primitive human relatives persisted millions of years longer than previously believed, outlasting even the Neanderthals. In 2008, researchers found bones of dwarf people thought to be modern humans in caves on the Pacific islands of Palau, but these remains also have small heads and some features considered primitive for modern humans. With this latest evidence, some now argue that these exceptionally small fossil remains are examples of *insular dwarfism*, an evolution toward smaller size that also happened to mammals like elephants and extinct mammoths living on islands. See Maggie Fox, "Tiny Palau Skeletons Suggest 'Hobbits' Were Dwarfs," Reuters, Mar. 10, 2008, http://africa.reuters.com/wire/news/usnN10595112.html.

108 *Wielding fire:* Johan Goudsblom, *Fire and Civilization* (London: Penguin, 1994), 29–30.

109 *a big brain is very expensive:* John Morgan Allman, *Evolving Brains* (New York: Scientific American Library, 1999).

109 *compete with other organs:* Leslie C. Aiello and Peter Wheeler, "The Expensive-Tissue Hypothesis: The Brain and the Digestive System in Human and Primate Evolution," *Current Anthropology* 35 (1995): 199–221.

109 *Heating food has several benefits:* Goudsblom, *Fire and Civilization*, 34–35.

109 *20 percent of the body's energy:* Potts, *Humanity's*, 206.

109 *size increased by 24 percent:* Allman, 194.

109 *first true brains:* Jun-Yuan Chen, Di-Ying Huang, and Chia-Wei Li, "An Early Cambrian Craniate-like Chordate," *Nature* 402 (1999): 518–22, www .nature.com/nature/journal/v402/n6761/full/402518a0.html.

110 *extended families:* Allman, 203, argues that extended families were as critical as growing brain size. "The human evolutionary success story depends on two great buffers against misfortune, large brains and extended families, with each supporting and enhancing the adaptive value of the other."

110 *bright paints:* A number of recent discoveries suggest that humans and earlier ancestors may have used red ochre and other pigments far earlier, perhaps as long as 200,000 years ago. See Jonathan Amos, "A Colourful Beginning for Humanity," *BBC News*, Sept. 10, 2006, http://news.bbc.co.uk/2/hi/science/nature/5329486.stm.

110 *a single, fixed nature:* Paul R. Ehrlich, *Human Natures: Genes, Culture, and the Human Prospect* (Washington, D.C.: Island Press, 2000).

110 *"genetically specified behavioral predilections":* This kind of evolutionary argument is typified by Jerome H. Barkow, Leda Cosmides, and John Tooby, *The Adapted Mind* (New York: Oxford University Press, 1992).

111 *the perennial nature/nuture debate:* Peter J. Richerson and Robert Boyd, "Culture Is Part of Human Biology," in *Science Studies: Probing the Dynamics of Scientific Knowledge*, Sabine Maasen and Mattias Winterhager, eds. (Bielefeld, Germany: Transcript Verlag, 2001).

111 *against the grain of human instinct:* Robert Wright, *The Moral Animal* (New York: Pantheon, 1994), 94–96.

111 *the emerging field of cultural biology:* Steven R. Quartz and Terrence J. Sejnowski, *Liars, Lovers, and Heroes* (New York: HarperCollins, 2002).

112 *a shared system of meaning:* Mary E. Clark, *In Search of Human Nature* (New York: Routledge, 2002), 183–84, 230–32.

115 *universalism:* This view is exemplified by Francis Fukuyama, *The End of History and the Last Man* (New York: Free Press, 1992).

116 *"last great Enlightenment regime":* John Gray, *False Dawn* (New York: New Press, 1998), 2.

116 *hardwired according to a genetic blueprint:* Steven Pinker, *The Language Instinct* (New York: Morrow, 1994). Pinker compares a mind to a Swiss Army knife, (wired by genes) composed of modules that are adapted to various tasks.

116 *new field of cultural psychology:* This section relies on Richard E. Nisbett, *The Geography of Thought* (New York: Free Press, 2003).

118 *the way Japanese and American mothers teach:* Anne Fernald and Hiromi Morikawa, "Common Themes and Cultural Variations in Japanese and American Mothers' Speech to Infants," *Child Development* 64 (1993): 637–56, is discussed in Nisbett, *Geography*, 150.

118 *strong dose of Western ideas:* Judith Shapiro, *Mao's War Against Nature* (New York: Cambridge University Press, 2001), 9–11.

122 *daring pioneers:* Clive Gamble, *Timewalkers: The Prehistory of Global Colonization* (Cambridge, MA: Harvard University Press, 1993), 8–9.

122 *cultural sophistication:* Gamble, *Timewalkers*, 118–19.

123 *Kostenki settlement:* Ibid., 185–87.

123 *re-creations of mammoth-bone houses:* This website—http://donsmaps .com/mammothcamp.html—shows reconstructions of mammoth-bone dwellings.

123 *large oval longhouses:* Ian Shaw and Robert Jameson, eds., *A Dictionary of Archaeology* (Oxford: Blackwell Publishers, 1999), 342.

124 *long-distance trade:* Richard G. Klein, *Man and Culture in the Late Pleistocene* (San Francisco: Chandler Pub. Co., 1969), 227–28.

125 *Viking colony on Greenland:* The account of climate history and the Norse colonization relies on Brian Fagan, *The Little Ice Age* (New York: Basic Books, 2000).

126 *"They lived on the edge":* Buckland is quoted in Heather Pringle, "Death in Norse Greenland," *Science* 275 (1997): 924–26.

126 *The archaeological evidence:* This section on the Greenland colony draws from: Sophia Perdikaris and Thomas H. McGovern, "Cod Fish, Walrus, and Chieftains: Economic Intensification in the Norse North Atlantic," in *Seeking a Richer Harvest*, Tina L. Thurston and Christopher T. Fisher, eds. (New York: Springer, 2007), 193–216; Thomas H. McGovern and Sophia Perdikaris, "The Vikings' Silent Saga," *Natural History*, Oct. 2000, www.nhmag.com/features/ 1000_vikings.html; Thomas H. McGovern, "Cows, Harp Seals, and Churchbells: Adaptation and Extinction in Norse Greenland," *Human Ecology* 8 (1980): 245– 75; Andrew J. Dugmore, Christian Keller, and Thomas H. McGovern, "Have We Been Here Before? Climate Change and the Contrasting Fates of Human Settlements in the Atlantic Islands," paper presented at Human Security and Climate Change, an international workshop in Oslo, Norway, June 21–23, 2005, www.gechs.org/downloads/holmen/Dugmore_etal.pdf; and Dale Mackenzie Brown, "The Fate of Greenland's Vikings," *Archaeology*, Feb. 28, 2000, http:// www.archaeology.org/online/features/greenland/.

126 *"Any people not willing"*: Paul Bohannan, *How Culture Works* (New York: Free Press, 1995), 125, 173.

CHAPTER 6.   PLAYING PROSPERO:
THE TEMPTATIONS OF TECHNOFIX

130 *"the urge to control nature"*: René Dubos, *The Dreams of Reason* (New York: Columbia University Press, 1961), 57.

132 *"our relationship to the planet"*: Will Steffen, "Surviving the Anthropocene: The Great Challenges of the 21st Century," presentation given at the Stockholm Resilience Center, Oct. 30, 2007, http://www.stockholmresilience.org/seminar andevents/seminarandeventvideos/willsteffenthegreatchallengesofthe21stcentury. 5.aeea46911a3127427980003219.html.

132 *the first and only solution*: David W. Keith, "Geoengineering the Climate: History and Prospect," *Annual Review of Energy and the Environment* 25 (2000): 254.

132 *speak in euphemisms*: Keith, "Geoengineering the Climate," 46.

133 *pointed out half a century ago*: Roger Revelle and Hans E. Suess, "Carbon Dioxide Exchange Between Atmosphere and Ocean and the Question of an Increase of Atmospheric CO_2 During the Past Decades," *Tellus* 9 (1957): 18–27.

134 *"the ultimate form of slavery"*: James Lovelock, *The Revenge of Gaia* (London: Penguin Books, 2006), 152.

135 *anything save human life*: Mary Midgley, *Science and Poetry* (New York: Routledge, 2001), 184.

135 *a living whole*: Lewis Thomas, *The Lives of a Cell* (New York: Viking, 1974), 145.

136 *"fringe entertainment"*: Gavin Schmidt, "Geo-engineering in Vogue," *RealClimate*, June 28, 2006, www.realclimate.org/index.php/archives/2006/06/ geo-engineering-in-vogue.

136 *the arsenal in warfare*: Spencer Weart, "Climate Modification Schemes," American Institute of Physics website, www.aip.org/history/climate/RainMake.htm.

136 *"a more important weapon than the atom bomb"*: William B. Meyer, "The Life and Times of U.S. Weather: What Can We Do About It?" *American Heritage*, June–July, 1986, 36–48, http://www.americanheritage.com/articles/ magazine/ah/1986/4/1986_4_38.shtml.

136 *2,600 cloud-seeding missions*: Keith, "Geoengineering the Climate," 253.

136 *international treaty*: Convention on the Prohibition of Military or Any Other Hostile Use of Environmental Modification Techniques, entered into

force Oct. 5, 1978. The text is available at http://fletcher.tufts.edu/multi/texts/ BH700.txt.

137 *sorcerer's apprentice:* Johann Wolfgang von Goethe's 1779 poem "Der Zauberlehrling." German and English versions are available at http://german .about.com/library/blgzauberl.htm.

137 *Rapid City:* "The 1972 Black Hills–Rapid City Flood Revisited," South Dakota Water Science Center, U.S. Geological Survey, http://sd.water.usgs.gov/ projects/1972flood/index.html.

137 *again on the rise:* The National Academy of Sciences in a 2003 study, *Critical Issues in Weather Modification Research*, recommended "a renewed commitment to advancing our knowledge of fundamental atmospheric processes that are central to the issues of intentional and inadvertent weather modification"; http:// books.nap.edu/catalog.php?record_id = 10829#toc. Members of Congress from rain-starved states in the West and Southwest are promoting expanded rainmaking programs.

137 *Beijing Olympics:* "Beijing Disperses Rain to Dry Olympic Night," Xinhua, Aug. 9, 2008, http://news.xinhuanet.com/english/2008-08/09/content_9079637 .htm.

137 *a scientific manifesto:* N. Rusin and L. Flit, "Man Versus Climate," trans. Dorian Rottenberg (Moscow: Peace Publishers, 1960), quoted in Stephen H. Schneider, "Geoengineering: Could—or Should—We Do It?" *Climatic Change* 33 (1996): 292.

138 *"We have described those mysteries":* Rusin and Flit, "Man Versus Climate," quoted in David W. Keith, "Geoengineering the Climate," 251.

138 *most enthusiastic advocate:* The account of Teller's career relies on Peter Goodchild, *Edward Teller: The Real Dr. Strangelove* (Cambridge, MA: Harvard University Press, 2004).

139 *"not worth worrying about":* Goodchild, *Edward Teller,* 275–76.

139 *a paper on geoengineering:* E. Teller, L. Wood, and R. Hyde, "Global Warming and Ice Ages: Prospects for Physics-Based Modulation of Global Change," Lawrence Livermore National Laboratory (1997): UCRL-JC-128715.

139 *an opinion piece:* Edward Teller, "Sunscreen for Planet Earth," *Wall Street Journal,* Oct. 17, 1997. This has been reprinted in the *Hoover Digest* and is available at www.hoover.org/publications/digest/3522851.html.

139 *paid tribute:* "Dr. Strangelove saves the earth," *The Economist,* Jan. 15, 2007. http://www.economist.com/world/international/displaystory.cfm?story_id = 8542549.

140 *Budyko:* Schneider, "Geoengineering: Could—or Should—We Do It?" 293.

141 *virtually stopped cold:* Oliver Morton, "Is This What It Takes to Save the World?" *Nature* 447 (2007): 132–36.

141 *recent scientific assessments:* Keith's "Geoengineering the Climate" provides a history of these assessments.

141 *"considered taboo":* Wallace S. Broecker, *How to Build a Habitable Planet* (Palisades, NY: Eldigio Press, 1985).

141 *came roaring out of the scientific closet:* Paul Crutzen, "Albedo Enhancement by Stratospheric Sulfur Injections: A Contribution to Resolve a Policy Dilemma," *Climatic Change* 77 (2006): 211–19.

141 *"environmental worrier":* Schneider is quoted in Morton, "Is This What It Takes?" 13.

142 *One recent modeling study:* Meinrat Andreae, Chris D. Jones, and Peter M. Cox, "Strong Present-Day Aerosol Cooling Implies a Hot Future," *Nature* 435 (2005): 1187–90. See as well the news story in same issue: Quirin Schiermier, "Clear Skies Raise Global-Warming Estimates": 1142–43.

143 *"like a junkie":* Andreae quoted in Morton, "Is This What It Takes?" 132.

143 *Ralph Cicerone:* William J. Broad, "How to Cool a Planet (Maybe)," *New York Times*, June 27, 2006, D4.

143 *consider a moratorium:* Ralph Cicerone, "Geoengineering: Encouraging Research and Overseeing Implementation," *Climatic Change* 77 (2006): 221–26.

143 *a provocative essay:* Brad Allenby, "Earth Systems Engineering and Management," *Technology and Society Magazine*, IEEE, Winter 2000–2001, 10–24.

144 *geoengineering could buy time:* T. M. L. Wigley, "A Combined Mitigation/Geoengineering Approach to Climate Stabilization," *Science* 13 (2006): 452–54.

145 *"not a single one":* Olive Heffernan, "Research Is Responsibility," *Nature Reports: Climate Change* 2 (May 2008).

145 *"exceptionally dangerous":* Will Steffen, "Will Technology Spare the Planet?" in *Challenges of a Changing Earth*, Will Steffen, et al., eds. (Heidelberg: Springer-Verlag, 2002), 189–91.

146 *1.8 percent of incoming:* Lee Lane, et al., "Workshop Report on Managing Solar Radiation," Nov. 2006 workshop sponsored by NASA Ames Research Center and the Carnegie Institution of Washington Department of Global Ecology at Stanford University on the use of "solar radiation management" for coping with climate change, NASA/CP-2007-214558.

146   *immense cloud of gas:* Kenneth A. McGee, et al., "Impacts of Volcanic Gases on Climate, the Environment, and People," U.S. Geological Survey OpenFile Report 97-262 (1997), http://pubs.usgs.gov/of/1997/of97-262/of97-262.html.

147   *Roger Angel:* Roger Angel, "Feasibility of Cooling the Earth with a Cloud of Small Spacecraft near the Inner Lagrange Point (L1)," *Proceedings of the National Academy of Sciences* 103 (2006): 17184–89.

147   *down-to-earth strategy:* Lane, et al., "Workshop Report," 6–8.

148   *Shell-forming plankton:* Graham Phillips, "Ocean Time Bomb," *Sidney Morning Herald,* Sept. 12, 2007, www.smh.com.au/news/environment/ocean-time-bomb/2007/09/11/1189276723526.html.

148   *eruptions affected rainfall:* Kevin E. Trenberth and Aiguo Dai, "Effects of Mount Pinatubo Volcanic Eruption on the Hydrological Cycle as an Analog of Geoengineering," *Geophysical Research Letters* 34 (2007): L15702; Luke Oman et al., "High-Latitude Eruptions Cast Shadow over the African Monsoon and the Flow of the Nile," *Geophysical Research Letters* 33 (2006): L18711.

148   *grim forecast of decreased rainfall:* H. Damon Matthews and Ken Caldeira, "Transient Climate-Carbon Simulations of Planetary Geoengineering," *Proceedings of the National Academy of Sciences* 104 (2007): 9949–54.

149   *five hundred years:* Lennart Bengtsson, "Geoengineering to Confine Climate Change: Is It at All Feasible?" *Climatic Change* 77 (2006): 229–34.

149   *This institutional challenge:* David Chandler, "A Sunshade for the Planet," *New Scientist,* July 21, 2007, http://environment.newscientist.com/channel/earth/mg19526131.700-a-sunshade-for-the-planet.html.

150   *interest in ocean fertilization:* Emma Young, "Can 'Fertilising' the Ocean Combat Climate Change?" *New Scientist,* Sept. 12, 2007, 42–45, http://environment.newscientist.com/channel/earth/mg19526210.600-can-fertilising-the-ocean-combat-climate-change.html.

152   *"not going to make a significant difference":* Steven Mufson, "Iron to Plankton to Carbon Credits," *Washington Post,* July 20, 2007, D01, www.washingtonpost.com/wp-dyn/content/article/2007/07/19/AR2007071902553.html.

152   *deplete oxygen levels:* Sallie W. Chisholm, Paul G. Falkowski, and John J. Cullen,"Dis-Crediting Ocean Fertilization," *Science* 294 (2001): 309–10.

152   *"Thousands of species depend":* Chisholm is quoted in Young, "Can 'Fertilising' . . . ?"

152   *a statement of concern:* This is available at www.imo.org/includes/blast Data.asp/doc_id = 8272/14.pdf.

153 *Planktos canceled:* Rachel Petkewich, "Fertilizing the Ocean with Iron," *Chemical & Engineering News*, Mar. 31, 2008, 30–33, http://pubs.acs.org/cen/science/86/8613sci1.html.

153 *Climos . . . was seeking permits:* Carrie Peyton Dahlberg, "S. F. Entrepreneur Floats a Bold Idea to 'Fertilize' Ocean," *Sacramento Bee*, Mar. 30, 2008, www.sacbee.com/101/story/821934.html.

153 *called for a moratorium:* Jeff Tollefson, "UN Decision Puts Brakes on Ocean Fertilization," *Nature* 453 (2008): 704.

153 *KlimaFa, made a gift:* Elisabeth Rosenthal, "Vatican Penance: Forgive Us Our Carbon Output," *New York Times*, Sept. 17, 2007, www.nytimes.com/2007/09/17/world/europe/17carbon.html.

154 *only a few percent:* Michael J. Lutz, et al., "Seasonal Rhythms of Net Primary Production and Particulate Organic Carbon Flux to Depth Describe the Efficiency of Biological Pump in the Global Ocean," *Journal of Geophysical Research* 112 (2007): C10011.

154 *as much as 98 percent:* Young, "Can 'Fertilising' . . . ?"

154 *more frequent forest fires:* The reports of this include: Sylvia Poggioli, "Blazes in Greece Wake-up Call for Climate Change," *All Things Considered*, National Public Radio, Sept. 4, 2007; Roger Highfield, "Siberian Forest Fires due to Climate Change," *Telegraph* online, Aug. 1, 2007, www.telegraph.co.uk/earth/main.jhtml?xml=/earth/2007/08/01/scisiberia101.xml; Dan Joling, "Warming Reshapes Alaska's Forest," Associated Press in *Jackson Hole Star Tribune*, Sept. 11, 2006, www.jacksonholestartrib.com/articles/2006/09/11/news/regional/17bac97812d32299872571e500210e84.txt; Tom Knudson, "Global Warming Fuels Hotter Western Fires," *Sacramento Bee*, Nov. 30, 2008, www.fresnobee.com/local/story/1043621.html.

154 *Schwarzenegger described: This Week with George Stephanopoulos*, ABC News, Nov. 16, 2008, http://abcnews.go.com/Video/playerIndex?id=6264603.

154 *large, devastating wildfires:* A. L. Westerling, et al., "Warming and Earlier Spring Increase Western U.S. Forest Wildfire Activity, *Science* 313 (2006): 940–43.

155 *forests may become a source:* Steven W. Running, "Is Global Warming Causing More, Larger Wildfires?" *Science* 313 (2006): 927–28, www.sciencemag.org/cgi/content/full/313/5789/927.

155 *38 percent:* Bette Hileman and Jeff Johnson, "Driving CO_2 Underground," *Chemical & Engineering News*, Sept. 24, 2007, 74–81.

155 *problem will possibly double:* Robert H. Socolow, "Can We Bury Global Warming?" *Scientific American*, July 2005, 49–55.

155 *"The climate problem . . . is a coal problem":* Jonathan Shaw, "Fueling Our Future," *Harvard Magazine*, May–June 2006, 45.

156 *only one power plant in the world:* Alok Jha, "New Era for Fossil Fuels as Carbon Capturing Power Plant Begins Work," *The Guardian*, April 9, 2009, www.guardian.co.uk/environment/2009/apr/08/first-carbon-capture-power-plant-lacq.

156 *designed for easy retrofit:* Simon Romero, "2 Industry Leaders Bet on Coal but Split on Cleaner Approach," *New York Times*, May 28, 2006, www .nytimes.com/2006/05/28/business/28coal.html.

156 *coal rush . . . began losing steam:* Mark Clayton, "Pace of Coal-Power Boom Slackens," *Christian Science Monitor*, Oct. 25, 2007, www.csmonitor.com/2007/1025/p03s02-wogi.html.

157 *utilities abandoned plans:* Steve James, "Coal's Time Is Up in US, Environmentalist Warns," Reuters, Feb. 15, 2008, www.planetark.com/dailynewsstory.cfm/newsid/46967/story.htm.

157 *Schellnhuber:* Roger Harrabin, "Climate Change Goal Unreachable," *BBC News*, Dec. 10, 2007, http://news.bbc.co.uk/2/hi/science/nature/7135836.stm.

157 *not as hard as one might imagine:* David Keith, "Iron Fertilization, Air Capture, and Geoengineering," presentation at the Woods Hole Oceanographic Institution's Symposium Program on Ocean Iron Fertilization, Woods Hole, MA, Sept. 26, 2007.

158 *first successful demonstration:* Wallace S. Broecker and Robert Kunzig, *Fixing Climate* (New York: Hill and Wang, 2008).

158 *Lackner estimates:* Alan Zarembo, "Solving Global Warming with Giant Vacuums," *Los Angeles Times*, Apr. 29, 2008, www.latimes.com/news/nation world/nation/la-sci-carbon29apr29,0,7995802,full.story.

158 *60 percent of annual CO_2:* Hileman and Johnson, "Driving CO_2 Underground."

159 *Developed countries:* Michael R. Raupach, et al., "Global and Regional Drivers of Accelerating CO_2 Emissions," *Proceedings of the National Academy of Sciences* 104 (2007): 10288–93.

159 *historical debt:* This data is available from the U.S. Department of Energy's Carbon Dioxide Information Analysis Center website: http://cdiac.ornl .gov/trends/emis/meth_reg.html. See also: Raupach, et al., "Global and Regional Drivers."

159   *test injection:* Richard A. Kerr, "A Possible Snag in Burying CO_2," *Science Now* Daily News, June 28, 2006, http://sciencenow.sciencemag.org/cgi/content/long/2006/628/3.

160   *To prevent such a return:* Wallace S. Broecker, "Deep-Sea Carbon Storage Must Be Tested, Says Leading Scientist," *The Guardian*, June 18, 2008, www.guardian.co.uk/environment/2008/jun/18/carboncapturestorage.carbonemissions.

160   *Initial experiments:* Peter G. Brewer, "Direct Injection of Carbon Dioxide into the Oceans" in *The Carbon Dioxide Dilemma: Promising Technologies and Policies*, National Research Council (Washington, D.C.: National Academies Press, 2003).

161   *beneath mud:* Kurt Zenz House, et al., "Permanent Carbon Dioxide Storage in Deep-Sea Sediments," *Proceedings of the National Academy of Sciences* 103 (2006): 12291–95.

161   *25 percent higher:* Alister Doyle, "Petrify, Liquefy: New Ways to Bury Greenhouse Gas," Reuters, May 7, 2008, www.reuters.com/article/markets News/idUSL0573285820080508.

161   *possible obstacles:* Michael Behar, "How Earth-Scale Engineering Can Save the Planet," *Popular Science*, June 2005, www.popsci.com/popsci/aviationspace/3afd8ca927d05010vgnvcm1000004eecbccdrcrd.html.

162   *One energy economist:* Ashok Gupta, the lead energy economist at the Natural Resources Defense Council, quoted in Amanda Griscom, "Tough Cell," *Grist*, Feb. 26, 2003, www.grist.org/news/powers/2003/02/26/tough/index.html.

162   *number of mountain rescues:* James Gorman, "The Call in the Wild," *New York Times*, Aug. 30, 2001, http://query.nytimes.com/gst/fullpage.html?res = 9C04E5DF1530F933A0575BC0A9679C8B63.

162   *development in these danger zones:* "Coastal Development and the National Flood Insurance Program," in *Oceans and Coastal Resources: A Briefing Book*, May 30, 1997, Congressional Research Service Report 97-588 ENR.

163   *"perfect moral storm":* Stephen M. Gardiner, "A Pefect Moral Storm: Climate Change, Intergenerational Ethics, and the Problem of Moral Corruption," *Environmental Values* 15 (2006) 397–413; and "Is Geoengineering the Lesser Evil?" *Environmental Research Letters*, Apr. 18, 2007, http://environmental researchweb.org/cws/article/opinion/27600.

163   *political hazards:* Edward A. Parson, "Reflections on Air Capture: The Political Economy of Active Intervention in the Global Environment," *Climatic Change* 74 (2006): 5–15.

163 *such claims are inevitable:* Stephen H. Schneider, "Earth Systems Engineering and Management," *Nature* 409 (2001): 417–21.

164 *failed miserably:* James Gustave Speth, *Red Sky at Morning* (New Haven: Yale University Press, 2004), 96.

165 *leading figures in Russia:* Alastair Gee, "Green Living: Russian Officials Put Upbeat Spin on Global Warming," Cox News Service, *Kansas City Star,* Aug. 5, 2007, D1.

165 *planted a flag on the ocean floor:* Lee Carter, "Arctic Neighbors Draw Up Battle Lines," *BBC News,* Aug. 11, 2007, http://news.bbc.co.uk/2/hi/americas/6941569.stm.

165 *"some kind of arms race":* Moises Velasquez-Manoff, "Scientists Weigh Risks of Climate 'Techno-fixes,' " *Christian Science Monitor,* Mar. 29, 2007, www.csmonitor.com/2007/0329/p13s02-sten.htm.

166 *a planet already in a state of emergency:* See "Humanity's Footprint 1961–2003," www.footprintnetwork.org/gfn_sub.php?content=global_footprint.Based on this accounting, the demands of human enterprise began to exceed the Earth's renewal capacity in the 1980s. It would require 1.3 Earths to meet current demands without depleting the planet's capacity to sustain life.

166 *a warning:* John Ruskin, "The Veins of Wealth," in *The Genius of John Ruskin,* John D. Rosenberg, ed. (Boston: Routledge & Kegan Paul, 1980), 244–53.

166 *In seeming despair:* Nicholas Georgescu-Roegen, "Energy and Economic Myths," *Southern Economic Journal* 41 (1975): 347–81.

167 *prophet Moses:* Deuteronomy 30:19.

CHAPTER 7. ON VULNERABILITY AND SURVIVABILITY

168 *"the power of mind":* Martin Luther King, "Antidotes to Fear," in *Strength of Love* (Philadelphia: Fortress Press, 1981), 118–19.

169 *three scientists:* Martin Parry, et al., "Squaring Up to Reality," *Nature Reports: Climate Change* 2 (2008): 68–70.

169 *burden on planetary systems:* Vaclav Smil, *Energy at the Crossroads* (Cambridge, MA: MIT Press, 2005), ch. 6.

170 *"index of far-reaching ruin":* John Ruskin, "The Veins of Wealth," in *The Genius of John Ruskin,* John D. Rosenberg, ed. (Boston: Routledge & Kegan Paul, 1980).

171 *breakdown of the global infrastructure:* Jad Mouawad and Julia Werdigier, "Warning on Impact of China and India Oil Demand," *New York Times,* Nov. 7, 2007, www.nytimes.com/2007/11/07/business/07cnd-energy.html, warns that

global energy security will be at increasing risk as the world relies primarily on the Middle East and Russia for oil.

172 *rise by at least 25 percent:* Jad Mouawad, "Cuts Urged in China's and India's Energy Growth," *New York Times,* Nov. 7, 2007, www.nytimes.com/2007/11/07/business/worldbusiness/07energy.html?scp=11&sq=&st=nyt.

172 *catastrophic consequences:* In late 2007, the IPCC issued the synthesis report from its fourth assessment, http://environment.newscientist.com/channel/earth/mg19626314.100-analysis-ipcc-issues-dire-climate-change-warning.html, which went further than its three detailed studies in warning that "warming could lead to some impacts that are abrupt or irreversible, depending upon the rate and magnitude of the climate change." This appeared to be a response to criticisms from scientists that the IPCC had underplayed the possible dangers.

172 *"we are twenty-five years too late":* David Biello, "State of the Science: Beyond the Worst Case Climate Change Scenario," *Scientific American,* Nov. 26, 2007, www.sciam.com/article.cfm?id=state-of-the-science-beyond-the-worst-climate-change-case.

172 *other kinds of problems:* "Global Risks 2007: A Global Risk Network Report," World Economic Forum, 2007, www.weforum.org/pdf/CSI/Global_Risks_2007.pdf.

172 *"tectonic stresses":* Thomas Homer-Dixon, *The Upside of Down* (Washington, DC: Island Press, 2006), 10–16.

173 *The modern way of life:* J. R. McNeill, *Something New Under the Sun: An Environmental History of the Twentieth-Century World* (New York: Norton, 2000), xxiii.

173 *fundamental vulnerability:* Brian Fagan, *The Long Summer* (New York: Basic Books, 2004).

175 *anthropological theories:* Mark Nathan Cohen, *Health and the Rise of Civilization* (New Haven: Yale University Press, 1989), 2–3.

175 *had to work harder:* Charles L. Redman, *Human Impact on Ancient Environments* (Tucson: University of Arizona Press, 1999), 160–64.

175 *abundant, reliable, and nutritious:* Cohen, *Health and the Rise of Civilization,* 96.

176 *Pierre Biard:* Le Père Pierre Biard, "Relation of New France, of Its Lands, Nature of the Country, and of Its Inhabitants," in *The Jesuit Relations and Allied Documents,* 1616, R. G. Thwaites, ed. (Cleveland: Burrows, 1897), 3: 84–85. Biard's fascinating account of Micmac life is is now available online at www.archive.org/details/jesuitrelation0304jesuuoft, making it far easier to read his description

of how they hunted and dined through the seasons: "In October and November comes the second hunt for elks and beavers; and then in December (wonderful providence of God) comes a fish called by them Pomano, which spawns under the ice." Bear, which Biard noted was "very good," appeared on the menu from February to mid-March.

176 *sixty miles a day:* The French have used several different *leagues* in the past. If Biard is using the typical seventeenth-century measurement, a league is equivalent to 1,666 toise, roughly 2 miles long. Biard reported that the Micmac could travel 30 to 40 leagues a day by canoe, which seems extraordinary. This figure may be for travel downriver from inland sites to the coast.

177 *"ultimate enabler":* Peter Bellwood, *First Farmers: The Origins of Agricultural Societies* (Oxford, UK: Blackwell, 2005), 20.

177 *megadroughts:* Brian Fagan, *The Great Warming* (New York: Bloomsbury Press, 2008).

177 *two ancient settlements:* The account of Abu Hureyra and Ur relies on Fagan's telling of their history in *The Long Summer.*

178 *a landscape rich in new opportunities:* This section relies on Fagan, *Long Summer,* and O. Bar-Yosef and R. H. Meadow, "The Origins of Agriculture in the Near East," in *Last Hunters, First Farmers: New Perspectives on the Prehistoric Transition to Agriculture,* T. Douglas Price and Anne Birgitte Gebauer, eds. (Santa Fe: School of American Research Press, 1995).

180 *people eventually returned:* Fagan, *Long Summer,* and Bellwood, *First Farmers.*

180 *city of Ur:* This account of Ur's rise and fall follows Fagan's history, *Long Summer,* 6–7, and ch. 7, "Drought and Cities."

182 *escalating vulnerability:* William H. McNeill, "Control and Catastrophe in Human Affairs," in *The Global Condition: Conquerors, Catastrophes, and Community* (Princeton: Princeton University Press, 1992), 135–49.

183 *his influential theory:* Joseph Tainter, *The Collapse of Complex Societies* (Cambridge, UK: Cambridge University Press, 1988).

183 *rapidly diminishing returns:* Marshall Sahlins, "The Original Affluent Society," in *Stone Age Economics,* 33, describes the "imminence of diminishing returns" as the economic motive for hunter-gatherers' mobile style of life. If hunter-gatherers stay put for more than a short time, they will have to go farther and farther to hunt and gather and thereby work harder to subsist.

183 *dilemma of diminishing returns:* Joseph Tainter, "Problem Solving: Complexity, History, Sustainability," *Population and Environment* 22 (2000), 3–40.

184   *For much of human history:* Kenneth Pomeranz and Steven Topik, *The World That Trade Created* (Armonk, NY: M. E. Sharpe, 1999).

184   *Some might argue:* Jan Aart Scholte, *Globalization: A Critical Introduction* (New York: Palgrave, 2000).

185   *unprecedented in human history:* Immanuel Wallerstein, *Utopistics* (New York: New Press, 1998), 9.

185   *"just-in-time":* Charles Atkinson, "McDonald's, a Guide to the Benefits of JIT," *Inventory Management Review*, Nov. 8, 2005, www.inventorymanage mentreview.org/justintime/index.html.

186   *three- to four-day supply:* "Threats," Section 5.4, "Import and Export," Agriculture and Agri-Food Canada, www4.agr.gc.ca/AAFC-AAC/display-afficher .do?id = 1184246543160&lang = e.

186   *dry grocery products:* Jay Coggins and Ben Senauer, "Grocery Retailing," in *U.S. Industry in 2000: Studies in Competitive Performance*, National Research Council (Washington, DC: National Academy Press, 1999), 165–68.

186   *sparked a blockade:* "UK Fuel Shortages Worsen," *BBC News*, Sept. 11, 2000, http://news.bbc.co.uk/2/hi/uk_news/919429.stm.

187   *faced paralysis:* David Strahan, *The Last Oil Shock* (London: John Murray, 2007), 13–16.

187   *imports cheaper milk:* Joseph E. Stiglitz, *Globalization and Its Discontents* (New York: Norton, 2003), 5. Stiglitz argues that while opening Jamaican markets to U.S. imports may have hurt local dairy farmers, it "meant poor children could get milk more cheaply."

187   *farmers in India:* Vandana Shiva, "At the Mercy of Globalisation," *Hindu*, Sept. 23, 2001, www.hindu.com/thehindu/2001/09/23/stories/13230614 .htm. Under the influence of globalisation, between 1991 and 2001, food production declined as Indian farmers shifted from growing grain to cotton and sugarcane for export.

187   *the world is losing:* See news accounts of ongoing losses: Elisabeth Rosenthal, "In Backyard Europe, Fading Biodiversity," *International Herald Tribune*, Nov. 5, 2007, http://iht.com/articles/2007/11/05/europe/seed.php; Fred Pearce, "World's Livestock Breeds Need Protection," *New Scientist*, Sept. 8, 2007, www .newscientist.com/channel/life/mg19526204.500-worlds-livestock-breeds-need -protection.html; Elisabeth Rosenthal, "In Poland, 'Green' Fields Besieged," *International Herald Tribune*, Apr. 3, 2008, www.iht.com/articles/2008/04/03/europe/ poland.php.

188   *"accident waiting to happen":* Lance H. Gunderson and C. S. Holling,

eds., *Panarchy: Understanding Transformation in Human and Natural Systems* (Washington, DC: Island Press, 2002), 45.

189   *life has endured:* Simon A. Levin, *Fragile Dominion* (Reading, MA: Perseus Books, 1999).

189   *make it more precarious:* Oran R. Young, et al., "The Globalization of Socio-ecological Systems: An Agenda for Scientific Research," *Global Environmental Change* 16 (2006): 304–16.

190   *Even routine disasters:* Barry C. Lynn, *End of the Line* (New York: Doubleday, 2005), 1–17.

191   *Global Risks 2007:* World Economic Forum, www.weforum.org/en/media/publications/GlobalRiskReports/index.htm.

191   *a medical Paul Revere:* The section on Osterholm relies on: Tim Gihring, "The Pandemic Prophecy," *Minnesota Monthly*, Apr. 2006, www.minnesotamonthly .com/media/Minnesota-Monthly/April-2006/The-Pandemic-Prophecy/index .php; Michael T. Osterholm, "A Weapon the World Needs," *Nature* 235 (2005): 417–18; Michael T. Osterholm, "Preparing for the Next Pandemic," *Foreign Affairs*, July–Aug. 2005, www.foreignaffairs.org/20050701faessay84402/michael-t-osterholm/preparing-for-the-next-pandemic.html.

192   *three quarters of the genetic diversity:* Elisabeth Rosenthal, "In Backyard Europe.

192   *a diverse gene pool:* "Globalization Threatens Farm Animal Gene Pool and Future Food Security, UN Warns," UN News Service, Dec. 15, 2006, www .un.org/apps/news/story.asp?NewsID = 20994#.

192   *Globalization of livestock markets:* Catherine Brahic, " 'Livestock Meltdown' Threatens Developing World," *New Scientist*, Sept. 4, 2007, http://environment.newscientist.com/channel/earth/dn12584-livestock-meltdown -threatens-developing-world.html.

193   *narrowing of the gene pool:* Peter Aldhous, "Humans Take Control of Evolution," *New Scientist*, Feb. 15, 2007, 6–7.

194   *such a precipitating event:* Wallace S. Broecker and Robert Kunzig, *Fixing Climate* (New York: Hill and Wang, 2008), ch. 11, "Megadroughts of the Past."

194   *sand dunes blanketed:* Mark Lynas, *Six Degrees: Our Future on a Hotter Planet* (Washington, DC: National Geographic, 2008), 30.

195   *Ed Cook:* Broecker and Kunzig, *Fixing Climate*, ch. 11.

195   *preventive or reactive:* Robert Repetto, "The Climate Crisis and the

Adaptation Myth," working paper no. 13, Yale School of Forestry and Environmental Studies, Fall 2008.

196 *The Dutch:* Alix Kroeger, "Dutch Pioneer Eco-homes." *BBC News*, Mar. 1, 2007, http://news.bbc.co.uk/go/pr/fr/-/2/hi/europe/6405359.stm; and Joe Palca, "In a Strategic Reversal, Dutch Embrace Floods," *Morning Edition*, National Public Radio, Jan. 22, 2008, www.npr.org/templates/story/story.php?storyId= 18229027.

197 *Emiliania huxleyi:* see www.soes.soton.ac.uk/staff/tt/.

198 *carbon to the deep sea:* See Helle Ploug, et al., "Production, Oxygen Respiration Rates, and Sinking Velocity of Copepod Fecal Pellets: Direct Measurements of Ballasting by Opal and Calcite," *Limnology and Oceanography* 53 (2008): 469–76. This study, looking at the fate of carbon-rich fecal pellets excreted by zooplankton, found that the pellets sank faster and carried much more carbon to the deep sea when the zooplankton fed on shell-forming algae such as *Ehux* compared to when they dined on plankton species that do not make shells.

198 *Gephyrocapsa:* Will Steffen, et al., *Global Change and the Earth System: A Planet Under Pressure* (Heidelberg: Springer-Verlag, 2004), 34.

201 *some retreat:* Larry Rohter, "Shipping Costs Start to Crimp Globalization," *New York Times*, Aug. 3, 2008, www.nytimes.com/2008/08/03/business/worldbusiness/03global.html?em.

201 *Jeffrey E. Garten:* quoted in Rohter, "Shipping Costs."

202 *"food sovereignty":* "Towards Food Sovereignty: Constructing an Alternative to the World Trade Organization's Agreement on Agriculture," a report by an international working group of civil society, farmer, and peasant organizations participating in a Farmers, Food and Trade International Workshop in Geneva, Switzerland, Feb. 19–21, 2003. This report explains food sovereignty, critiques free-trade proposals for agriculture, and proposes policy alternatives, www.nffc.net/Farmers%20Worldwide/FoodSovereignty_anAlternative.pdf.

202 *La Via Campesina:* More information about this organization can be found at its website, www.viacampesina.org/main_en/index.php?option=com_frontpage&Itemid=1.

203 *Commission on the Future of Food:* This group of leading thinkers and activists was initiated by the government of the Region of Tuscany, Italy, and by the Indian food-sovereignty activist Vandana Shiva.

203 *"Manifesto":* The document is available at the Region of Tuscany's website, www.arsia.toscana.it/petizione/clima.aspx.

204   *Global Crop Diversity Trust:* More on its activities can be found at www
.croptrust.org/main/mission.php.

204   *"climate-proof" varieties:* Mark Kinver, " 'Climate-proof' Crop Hunt Begins," *BBC News*, Sept. 23, 2008, http://news.bbc.co.uk/2/hi/science/nature/
7622920.stm.

204   *"doomsday vault":* Jacqueline Ruttimann, "Doomsday Food Store Takes
Pole Position," *Nature* 441 (2006): 912–13.

207   *precariousness of the world oil supply:* Strahan, *The Last Oil Shock*, 57–58.

207   *excessive speculation:* Michael W. Masters, testimony before the United
States Senate Committee on Homeland Security and Government Affairs, June
24, 2008.

208   *protests and food riots:* John Vidal, "Global Food Crisis Looms as Climate
Change and Fuel Shortages Bite," *Guardian*, Nov. 3, 2007, www.guardian.co.uk/
environment/2007/nov/03/food.climatechange.

208   *price surges:* David Hackett Fischer, *The Great Wave* (New York: Oxford
University Press, 1996), 256.

208   *the lack of reserves:* See the following news accounts: Sue Kirchhoff,
"Surplus U.S. Food Supplies Dry Up," *USA Today*, June 2, 2008, www.usatoday
.com/money/industries/food/2008-05-01-usda-food-supply_n.htm; David Ben-
nett, "Growers and Economists Push for Strategic Grain Reserves," *Delta Farm
Press*, Aug. 4, 2008, http://deltafarmpress.com/corn/grain-forum-0804/; "NFFC
Teleconference Outlines How Lack of Reserves Fuels Global Food Crisis," *High
Plains Journal*, July 30, 2008, www.hpj.com/archives/2008/aug08/aug4/NFFCtele
conferenceoutlinesh.cfm.

209   *"widest ranging market failure":* Alison Benjamin, "Stern: Climate Change
a 'Market Failure,' " *Guardian*, Nov. 29, 2007, www.guardian.co.uk/environment/
2007/nov/29/climatechange.carbonemissions.

210   *"tight chains":* Mark Buchanan, "This Economy Does Not Compute,"
New York Times, Oct. 1, 2008, www.nytimes.com/2008/10/01/opinion/01buchanan
.html.

210   *many different perspectives:* Will Steffen provides an insightful critique of
the current structure of knowledge in his talk, "Surviving the Anthropocene,"
www.stockholmresilience.org/program/src/home/seminarandevents/seminar
andeventvideos/willsteffenthegreatchallengesofthe21stcentury.5.a791285116833
497ab80009006.html.

211   *when it moves by leaps:* This section draws on many sources, in particular
Vaclav Havel, *The Art of the Impossible* (New York: Fromm International, 1998),

166; Fischer, *The Great Wave;* Immanuel Wallerstein, *Utopistics* (New York: New Press, 1998).

211   *Gabra:* David Maybury-Lewis, *Millennium: Tribal Wisdom and the Modern World* (New York: Viking, 1992), 81–85.

211   *Senka:* quoted in David Keys, *Catastrophe* (New York: Ballantine, 2000), 244.

212   *"a tragic epoch":* John Gray, *False Dawn* (New York: New Press, 1998), 207.

212   *"collective suicide":* Strahan, *The Last Oil Shock,* 212.

213   *"Do not try to plan the details":* C. S. Holling, "From Complex Regions to Complex World," *Ecology and Society* 9 (2004), www.ecologyandsociety.org./vol9/iss1/art11.

CHAPTER 8.   A NEW MAP FOR THE PLANETARY ERA

215   *Humans inevitably view nature:* Roy Rappaport, "Nature, Culture, and Ecological Anthropology," in *Man, Culture, and Society,* Harry L. Shapiro, ed. (London: Oxford University Press, 1971), 237–67.

216   *compares philosophy to plumbing:* Liz Else, "Mary, Mary quite contrary," *New Scientist,* November 03, 2001, 48, http://www.newscientist.com/article/mg17223154.800-mary-mary-quite-contrary.html.

217   *cultural autism:* Thomas Berry, "Ethics and Ecology" a paper delivered to the Harvard Seminar on Environmental Values, Cambridge, MA, April 9, 1996, http://www.earth-community.org/images/EthicsEcology1996.pdf.

217   *at odds with the way:* Kenneth Pomeranz and Steven Topik, *The World that Trade Created* (Armonk, N.Y.: M.E. Sharpe, 1999). The authors discuss in Chapter One the making of market conventions and how long it took to replace traditions of reciprocity, fair exchange, and a moral economy with the concept of prices established by supply and demand.

217   *Mauss:* Marcel Mauss, *The Gift* (New York: Norton, 1990), 76.

218   *an elder of the Weyewan:* David Maybury-Lewis, *Millennium: Tribal Wisdom and the Modern World* (New York: Viking, 1992), 72.

218   *aggressive intensity:* J. R. Ravetz, *The Merger of Knowledge with Power* (New York: Mansell Publishing, 1990), 97.

219   *implicit assumptions:* Arthur O. Lovejoy, *The Great Chain of Being* (New York: Harper & Row, 1936, 1960), 7.

219   *nomadic Gabra:* Maybury-Lewis, *Millennium,* 44.

219   *oft-repeated myths:* Theodore Roszak, "What a Piece of Work Is Man: Humanism, Religion, and the New Cosmology," *Network: The Scientific and Medical*

Network Review, Dec. 1999, 3–5, www.scimednet.org/Articles/RProszak.htm. This folklore about how the Scientific Revolution dethroned human arrogance has been recounted by Stephen Jay Gould in *Full House* and Carl Sagan in *The Demon-Haunted World*.

219 *exactly the opposite:* Lovejoy, *The Great Chain of Being*, 101–2.

220 *physically depicted:* Richard Tarnas, *The Passion of the Western Mind: Understanding the Ideas That Shaped Our World View* (New York: Ballantine, 1991), 195.

220 *The aim of life:* Basil Willey, *The Seventeenth Century Background* (New York: Doubleday, 1953), 39.

220 *Montaigne could rhapsodize:* Michel de Montaigne, "Of the Education of Children," quoted in Willey, *Seventeenth Century*, p. 41.

221 *medieval Scholastic philosophers:* Willey, *Seventeenth Century*, 23.

221 *"Heroes or Supermen":* Francis Bacon, "The Masculine Birth of Time," in *The Philosophy of Francis Bacon*, Benjamin Farrington, ed. (Chicago: University of Chicago Press, 1964), 72.

221 *a larger redemptive mission:* David Noble, *The Religion of Technology: The Divinity of Man and the Spirit of Invention* (New York: Knopf, 1997), ch. 4.

222 *"animating and controlling idea":* J. B. Bury, *The Idea of Progress* (New York: Dover, 1955). See Bury's preface in the Temple of Earth edition, 3a, for quotation cited, www.templeofearth.com/library.html.

222 *modern era's origin myth:* See Mircea Eliade, *Myth and Reality* (New York: Harper & Row, 1963); and Leo Marx, "The Domination of Nature and the Redefinition of Progress," in *Progress: Fact or Illusion*, Leo Marx and Bruce Mazlish, eds. (Ann Arbor: University of Michigan Press, 1998).

223 *"from death itself":* Charles Webster, *The Great Instauration* (New York: Holmes & Meier, 1976), 246.

223 *"price of progress is trouble":* Zay Jeffries, *Charles Franklin Kettering: 1876–1958* (Washington, DC: National Academy of Sciences, 1960), 113.

223 *"belief in inevitable progress":* Mary Midgley, *Evolution as a Religion: Strange Hopes and Stranger Fears* (London: Methuen, 1985), 81.

224 *"a living animal":* Carolyn Merchant, *The Death of Nature: Women, Ecology, and the Scientific Revolution* (New York: HarperCollins, 1980), ch. 4, p. 104.

224 *"new commercial empires":* Peter J. Bowler, *The Norton History of the Environmental Sciences* (New York: Norton, 1992), 69, 76.

225 *"veneration":* Robert Boyle, *A Free Enquiry into the Vulgarly Receiv'd Notion of Nature* (London: 1686), 18–19, quoted in Bowler, *Norton History*, 89.

225   *in some sense an illusion:* Paul Feyerabend, *The Conquest of Abundance: A Tale of Abstraction versus the Richness of Being* (Chicago: University of Chicago Press, 1999).

225   *unqualified materialism:* Boyle's 1666 work *Origin of Forms and Qualities* argues for mechanical philosophy and atomism. See Bowler, *Norton History*, 92.

226   *"totally devoid of mind":* René Descartes, *Principles of Philosophy*, in *Oeuvres Philosophiques*, ed. Alquié, 502, quoted in Mary Midgley, *Science and Poetry* (New York: Routledge, 2001), 42.

226   *animals do not feel pain:* See Descartes, *Oeuvres*, 3: 47–49 (letter 186, 1 April 1640), 85 (letter 192, 11 June 1640) quoted in Anita Guerrini, "The Ethics of Animal Experimentation in Seventeenth-Century England," *Journal of the History of Ideas* 50 (1989) 392. See also Keith Thomas, *Man and the Natural World: A History of the Modern Sensibility* (New York: Pantheon, 1983), 33.

226   *"Descartes still rules":* Midgley, *Science and Poetry*, 22.

227   *a good part of science:* See Richard Levins and Richard Lewontin, *The Dialectical Biologist* (Cambridge, MA: Harvard University Press, 1985), 1–2; Robert E. Ulanowicz, *Ecology, the Ascendent Perspective* (New York: Columbia University Press, 1997); and René Dubos, *The Dreams of Reason* (New York: Columbia University Press, 1961), 105.

227   *life's secrets:* Evelyn Fox Keller, *The Century of the Gene* (Cambridge, MA: Harvard University Press, 2000), 5–8, ch. 4, "Limits of Genetic Analysis: What Keeps Development on Track?"

227   *"mortal gods":* Francis Bacon, "The Refutation of Philosophies," in *The Philosophy of Francis Bacon*, Benjamin Farrington, ed. (Chicago: University of Chicago Press, 1964), 106.

228   *"to conquer and subdue":* Francis Bacon, "Thoughts and Conclusions" in *The Philosophy*, ed. Farrington, 93.

228   *breaking them down:* Merchant, *Death of Nature*, 182.

228   *"make her your slave":* Bacon, "The Masculine Birth of Time," in *The Philosophy*, ed. Farrington, 62.

228   *dream of human control:* Midgley, *Science and Poetry*, 25.

229   *a powerful tool:* Ilya Prigogine and Isabelle Stengers, *Order out of Chaos: Man's New Dialogue with Nature* (New York: Bantam, 1984), ch. 1.

229   *"obscure the fact":* Paul Davies and John Gribbin, *The Matter Myth* (New York: Simon & Schuster, 1992), 45–46.

229   *the limits of conventional analysis:* Ulanowicz, *Ecology*, ch. 1.

229   *challenges arising within science:* See Prigogine and Stengers, *Order out of Chaos*, ch. 1, "The Challenge to Science."

230   *hexagonal patterns:* David F. Peat, *From Certainty to Uncertainty* (Washington, D.C.: John Henry Press, 2002), 136.

230   *far-from-equilibrium systems:* See Prigogine and Stengers, *Order out of Chaos*, 171–176.

230   *can only be understood in their entirety:* Davies and Gribbin, *The Matter Myth*, 46–48.

231   *infelicitous but scientifically respectable:* Lawrence E. Joseph, *Gaia: The Growth of an Idea* (New York: St. Martin's Press, 1990), 29.

232   *more than merely "adapt":* Timothy M. Lenton, "Gaia and Natural Selection," *Nature* 394 (1998): 439–47.

232   *life as a planetary phenomenon:* Lynn Margulis and Dorion Sagan, *Slanted Truths* (New York: Springer-Verlag, 1997), 154.

232   *perhaps 50 billion species:* David M. Raup, *Extinction: Bad Genes or Bad Luck?* (New York: Norton, 1991), 3–4.

232   *imaginative vision:* Midgley, *Science and Poetry*.

232   *"All great theories are expansive":* Stephen Jay Gould, *Time's Arrow, Time's Cycle* (Cambridge, MA: Harvard University Press, 1987), 9.

233   *got its best reception:* Crispin Tickell, "Memoirs of a Reluctant Cult Figure," review of *Homage to Gaia*, by James Lovelock, *Nature* 409 (2001): 453–54.

233   *"Profoundly erroneous":* Joseph, *Gaia: The Growth of an Idea*, 56.

234   *different from Darwinian natural selection:* Lenton, "Gaia and Natural Selection," 442.

234   *"sequential selection":* Richard A. Betts and Timothy M. Lenton, "Second Chance for Lucky Gaia: A Hypothesis of Sequential Selection," *Gaia Circular*, The Geological Society of London (2007): 4–6.

235   *Others theorize:* Peter Westbroek, "Let's Reclaim Gaia for Science," *Palaeontologia Electronica* (2000), no. 1, http://palaeo-electronica.org/2000_1/editor/westbroe.htm.

235   *whether Earth is a living entity:* Lewis Thomas, "The Lives of a Cell" and "The World's Biggest Membrane," in *The Lives of a Cell* (New York: Viking, 1974).

235   *"the kind of hypothesis":* Bowler, *Norton History*, 517.

236   *"deprives us":* Stephen Toulmin, *The Return to Cosmology* (Berkeley: University of California Press, 1982). See p. 7 and the discussion of disciplinary fragmentation in part 3.

236 *"a term of abuse":* Mary Midgley, "Gaia: The Next Big Idea" (London: Demos, 2001), 14, www.demos.co.uk/publications/gaia.

236 *one scientist was warned:* Jim Gillon, "Feedback on Gaia," *Nature* 406 (2000): 685–86.

236 *"a brilliant organizing principle":* Joseph, *Gaia*, 9–10.

237 *to launch the Gaia society:* "Scientists Put Their Weight Behind Gaia Theory," *Nature* 391 (1998): 627, www.nature.com/nature/journal/v391/n6668/full/391626a0.html.

237 *Amsterdam Declaration:* The text is available at www.grida.no/news/press/2187.aspx?p = 2.

237 *The Gaian framework:* Jon Turney, *Lovelock and Gaia* (New York: Columbia University Press, 2003), 140–42.

237 *the all-mother of Greek mythology:* J. Donald Hughes, "GAIA: An Ancient View of our Planet," *Ecologist* 13 (1983), 54–60, describes three classical conceptions of the Earth: Earth as goddess, Earth as a living organism, and Earth in a reciprocal relationship with humans.

238 *the dancing Shiva:* Prigogine and Stengers propose the dancing Shiva as a new metaphor for nature at the end of the first chapter of *Order out of Chaos*.

238 *an impassioned plea:* Fred Pearce, "Gaia, Gaia: Don't Go Away," *New Scientist*, May 28, 1994, 43–44.

238 *the extended metaphor of religion:* Gregory Bateson and Mary Catherine Bateson, *Angels Fear: Towards an Epistemology of the Sacred* (New York: Macmillan, 1987), 200.

239 *"this lost integrity":* Vaclav Havel, *The Art of the Impossible* (New York: Fromm International, 1998), 171.

240 *"our nonexistence as entities:"* Thomas, "The Lives of a Cell," in *Lives of a Cell*, 3.

240 *"Every individual":* Samuel Butler, *Life and Habit* (New York: Dutton, 1923; originally published 1878), 85.

240 *a vast population of microbes:* Steven R. Gill, et al., "Metagenomic Analysis of the Human Distal Gut Microbiome," *Science* 312 (2006): 1355–59.

240 *"Life did not take over":* Lynn Margulis, "Power to the Protoctists," in Margulis and Sagan, *Slanted Truths*, 78. See also Martin Novak, "Five Rules for the Evolution of Cooperation," *Science* 314 (2006): 1560–63. Novak concludes that "cooperation is the secret behind the open-endedness of the evolutionary process."

241 *the establishment answer:* Andrew Revkin, "Forget Nature. Even Eden Is Engineered," *New York Times*, Aug. 20, 2002.

242 *"partnership"*: Carolyn Merchant, *Reinventing Eden* (New York: Routledge, 2003), ch. 11.

243 *"square the circle"*: Jeffrey Sachs, "The New Geopolitics," *Scientific American*, June 2006, 30.

244 *"wholly unrealisable fantasy"*: John Gray, *Beyond the New Right* (New York: Routledge, 1993), 142.

244 *a stationary or steady state:* Questioning growth is becoming a bit more respectable, as evidenced by the Oct. 15, 2008, issue of the British magazine *New Scientist*, which featured a cover declaring "The Folly of Growth," an editorial, "Time to Banish the God of Growth," and a special section with articles by Herman Daly and other thinkers challenging conventional economic wisdom.

244 *homecoming:* See Toulmin, "The Fire and the Rose," in *Return to Cosmology*, 255–74.

245 *Alliance to Rescue Civilization:* Richard Morgan, "Life After Earth: Imagining Survival Beyond This Terra Firma," *New York Times*, Aug. 1, 2006.

245 *latest version of escapist dreams:* See Mary Midgley, *Science as Salvation* (New York: Routledge, 1992), 159–63; and Nobel, *Religion of Technology*, ch. 9.

245 *a news story about Hawking's proposal:* Lucy Sherriff, "Hawking: Leave Earth or Die!" *Register,* June 14, 2006, www.theregister.co.uk/2006/06/14/hawkings_leave_earth/.

246 *to the Persian prophet Zoroaster:* Norman Cohn, *Cosmos, Chaos, and the World to Come* (New Haven: Yale University Press, 1993), ch. 4.

Acknowledgments

Though a writer spends a lot of time alone while writing a book, thinking and writing are not an individual undertaking but rather inescapably part of a larger, collaborative cultural enterprise. Any one person's effort, especially one such as this, incurs countless debts that are impossible to fully acknowledge, for one enters into an ongoing conversation that stretches across centuries, even millennia, as well as across different human cultures and specialized fields of study. I have done my best to credit these many voices in my endnotes and pray that I have not overlooked any. It has been heartening to discover so many people with diverse perspectives bringing the best of their minds, learning, and experience to bear on questions that help to illuminate the planetary crisis shaping this century.

This book has been a long and daunting journey, which I would have never completed without more than a little help from my friends and the kindness of strangers. I owe a special debt to Ross Gelbspan, a longtime colleague, friend, and fellow pilgrim in search of the human future. Over countless cups of coffee, he suffered with me through the vicissitudes of this seemingly impossible project; brought his passion, intellect, and expansive humanity to my far-ranging exploration; and patiently read draft after draft as I struggled to wrestle the many threads of thought and intuition into coherence. Most important, he maintained a steadfast faith that there was a promising seed buried in my early meanderings, a seed that ultimately did find the light and grow into this book. Nobody did more to help me through the lowest moments along the way than my former college roommate, Joan Doran Hedrick, who offered not only moral support but thoughtful advice about the possible lessons in a setback and about how to recover and forge ahead. Another friend, Marty Fujita, has been an irrepressible cheerleader-in-chief, driving me onward with her demand to "get this book done."

I once again want to thank friends and colleagues who engaged in or, I fear, sometimes simply endured brainstorming sessions that allowed me to air my thinking and struggle for the words to describe this new historical landscape, chief among them Len Ackland, Teresa Allen, Carol Stocker, Dolores Kong, Shelly Krimsky, and Ted Schettler. Their patience, interest, and indulgence were especially important in the beginning, when it wasn't at all clear where this effort was headed. The Center for Environmental Journalism at the University of Colorado and the Mesa Refuge for writers in Point Reyes Station, California, also offered support at important points early in this undertaking.

Later on, others already hard at work in addressing this planetary emergency made time in their own busy lives to read the manuscript and often improve it through their suggestions. These include Tony Cortese, Mary Evelyn Tucker, Linda Harrar, Molly Anderson, Randy Hayes, and Will Steffen.

I have been fortunate, indeed, to have my agents Joe Spieler and Deirdre Mullane, who appreciated the unusual nature of this book and did their utmost to get it published. No writer could ask for more astute and conscientious editors than Rachel Klayman and Lucinda Bartley, who put their hearts and minds into shepherding this book and have done so much to make it better.

But I owe the greatest debt of all to my husband, Carlo Obligato, who urged me to give up paid work so I could devote my full energy to this project and develop it into a book proposal. He has been not only a patient listener and perceptive reader but also an enthusiastic patron—my own Cosimo de' Medici on a public defender's salary, who sustained our household and never so much as hinted that he missed the modest pleasures we could no longer afford. His generosity of spirit has enriched my life immeasurably. For this and more, he has my abiding love.

Finally, there are three others I would like to remember—S., B., and D.—for their vital wordless contributions, for they made days in hermitlike isolation bearable. Unlike Ulysses' long-lived companion, two of them did not live to see the end of this odyssey.

Index